Lecture Notes in Computer Scien

T0238022

Commenced Publication in 1973
Founding and Former Series Editors:
Gerhard Goos, Juris Hartmanis, and Jan van Leeuwen

Editorial Board

David Hutchison
  Lancaster University, UK
Takeo Kanade
  Carnegie Mellon University, Pittsburgh, PA, USA
Josef Kittler
  University of Surrey, Guildford, UK
Jon M. Kleinberg
  Cornell University, Ithaca, NY, USA
Friedemann Mattern
  ETH Zurich, Switzerland
John C. Mitchell
  Stanford University, CA, USA
Moni Naor
  Weizmann Institute of Science, Rehovot, Israel
Oscar Nierstrasz
  University of Bern, Switzerland
C. Pandu Rangan
  Indian Institute of Technology, Madras, India
Bernhard Steffen
  University of Dortmund, Germany
Madhu Sudan
  Massachusetts Institute of Technology, MA, USA
Demetri Terzopoulos
  University of California, Los Angeles, CA, USA
Doug Tygar
  University of California, Berkeley, CA, USA
Moshe Y. Vardi
  Rice University, Houston, TX, USA
Gerhard Weikum
  Max-Planck Institute of Computer Science, Saarbruecken, Germany

Juraj Hromkovič   Richard Královič
Marc Nunkesser   Peter Widmayer (Eds.)

Stochastic Algorithms: Foundations and Applications

4th International Symposium, SAGA 2007
Zurich, Switzerland, September 13-14, 2007
Proceedings

 Springer

Volume Editors

Juraj Hromkovič
Richard Královič
Marc Nunkesser
Peter Widmayer
Swiss Federal Institute of Technology
Department of Computer Science ETH Zentrum
8092 Zürich, Switzerland
E-mail: {juraj.hromkovic,richard.kralovic,mnunkess,widmayer}@inf.ethz.ch

Library of Congress Control Number: 2007934297

CR Subject Classification (1998): F.2, F.1.2, G.1.2, G.1.6, G.2, G.3

LNCS Sublibrary: SL 1 – Theoretical Computer Science and General Issues

ISSN	0302-9743
ISBN-10	3-540-74870-9 Springer Berlin Heidelberg New York
ISBN-13	978-3-540-74870-0 Springer Berlin Heidelberg New York

This work is subject to copyright. All rights are reserved, whether the whole or part of the material is concerned, specifically the rights of translation, reprinting, re-use of illustrations, recitation, broadcasting, reproduction on microfilms or in any other way, and storage in data banks. Duplication of this publication or parts thereof is permitted only under the provisions of the German Copyright Law of September 9, 1965, in its current version, and permission for use must always be obtained from Springer. Violations are liable to prosecution under the German Copyright Law.

Springer is a part of Springer Science+Business Media

springer.com

© Springer-Verlag Berlin Heidelberg 2007
Printed in Germany

Typesetting: Camera-ready by author, data conversion by Scientific Publishing Services, Chennai, India
Printed on acid-free paper   SPIN: 12121038   06/3180   5 4 3 2 1 0

Preface

The 4th Symposium on Stochastic Algorithms, Foundations and Applications (SAGA 2007) took place on September 13–14, 2007, in Zürich (Switzerland). It offered the opportunity to present original research on the design and analysis of randomized algorithms, complexity theory of randomized computations, random combinatorial structures, implementation, experimental evaluation and real-world application of stochastic algorithms. In particular, the focus of the SAGA symposia series is on investigating the power of randomization in algorithmics, and on the theory of stochastic processes especially within realistic scenarios and applications. Thus, the scope of the symposium ranges from the study of theoretical fundamentals of randomized computation to experimental algorithmics related to stochastic approaches.

Previous SAGA symposia took place in Berlin (2001), Hatfield (2003), and Moscow (2005). This year 31 submissions were received, and the program committee selected 9 submissions for presentation. All papers were evaluated by at least four members of the program committee, partly with the assistance of subreferees. We thank the members of the program committee as well as all subreferees for their thorough and careful work. A special thanks goes to Harry Buhrman, Martin Dietzfelbinger, Rūsiņš Freivalds, Paul G. Spirakis, and Aravind Srinivasan, who accepted our invitation to give invited talks at SAGA 2007 and so to share their insights on new developments in research areas of key interest.

September 2007

Juraj Hromkovič
Richard Královič
Marc Nunkesser
Peter Widmayer

Organization

Program Chairs

Juraj Hromkovič
Peter Widmayer

Program Committee

Farid Ablayev
Dimitris Achlioptas
Andreas A. Albrecht
Andris Ambainis
Eli Ben-Sasson
Markus Bläser
Harry Buhrman
Colin Cooper
Josep Diaz
Martin Dietzfelbinger
Oktay M. Kasim-Zade
C. Pandu Rangan
Vijaya Ramachandran
Jose Rolim
Vishal Sanwalani
Martin Sauerhoff
Christian Scheideler
Georg Schnitger
Jiri Sgall
Angelika Steger
Kathleen Steinhöfel
Berthold Vöcking
Osamu Watanabe

Local Organization

Richard Královič
Tobias Mömke
Marc Nunkesser

External Reviewers

Peter Bosman
Alex Fukunaga
Aida Gainutdinova
Peter vd Gulik
Florian Jug
Fabian Kuhn
Julian Lorenz
Conrado Martinez
Dieter Mitsche
Jan Remy
Robert Spalek
Dirk Sudholt
Falk Unger
Nikolai Vereshchagin
Andrey Voronenko

Table of Contents

Invited Papers

On Computation and Communication with Small Bias 1
   Harry Buhrman

Design Strategies for Minimal Perfect Hash Functions 2
   Martin Dietzfelbinger

Hamming, Permutations and Automata 18
   Rūsiņš Freivalds

Probabilistic Techniques in Algorithmic Game Theory 30
   Spyros C. Kontogiannis and Paul G. Spirakis

Randomized Algorithms and Probabilistic Analysis in Wireless
Networking 54
   Aravind Srinivasan

Contributed Papers

A First Step Towards Analyzing the Convergence Time in
Player-Specific Singleton Congestion Games 58
   Heiner Ackermann

Communication Problems in Random Line-of-Sight Ad-Hoc Radio
Networks 70
   Artur Czumaj and Xin Wang

Approximate Discovery of Random Graphs 82
   Thomas Erlebach, Alexander Hall, and Matúš Mihal'ák

A VNS Algorithm for Noisy Problems and Its Application to Project
Portfolio Analysis 93
   Walter J. Gutjahr, Stefan Katzensteiner, and Peter Reiter

Digit Set Randomization in Elliptic Curve Cryptography 105
   David Jao, S. Ramesh Raju, and Ramarathnam Venkatesan

Lower Bounds for Hit-and-Run Direct Search 118
   Jens Jägersküpper

An Exponential Gap Between LasVegas and Deterministic Sweeping
Finite Automata 130
   Christos Kapoutsis, Richard Královič, and Tobias Mömke

Stochastic Methods for Dynamic OVSF Code Assignment in 3G
Networks 142
    Mustafa Karakoc and Adnan Kavak

On the Support Size of Stable Strategies in Random Games 154
    Spyros C. Kontogiannis and Paul G. Spirakis

Author Index 167

On Computation and Communication with Small Bias

Harry Buhrman

Centrum voor Wiskunde en Informatica (CWI) & University of Amsterdam
The Netherlands

Abstract. Many models in theoretical computer science allow for computations or representations where the answer is only slightly biased in the right direction. The best-known of these is the complexity class PP, for "probabilistic polynomial time". A language is in PP if there is a randomized polynomial-time Turing machine whose acceptance probability is greater than 1/2 if, and only if, its input is in the language.

Most computational complexity classes have an analogous class in communication complexity. The class PP in fact has two, a version with weakly restricted bias called PPcc, and a version with unrestricted bias called UPPcc. Ever since their introduction by Babai, Frankl, and Simon in 1986, it has been open whether these classes are the same. We show that PPcc is strictly included in UPPcc. Our proof combines a query complexity separation due to Beigel with a technique of Razborov that translates the acceptance probability of quantum protocols to polynomials. We will discuss some complexity theoretical consequences of this separation. This presentation is bases on joined work with Nikolay Vereshchagin and Ronald de Wolf.

J. Hromkovič et al. (Eds.): SAGA 2007, LNCS 4665, p. 1, 2007.
© Springer-Verlag Berlin Heidelberg 2007

Design Strategies for
Minimal Perfect Hash Functions

Martin Dietzfelbinger

Technische Universität Ilmenau, 98684 Ilmenau, Germany
martin.dietzfelbinger@tu-ilmenau.de

Abstract. A minimal perfect hash function h for a set $S \subseteq U$ of size n is a function $h: U \to \{0, \ldots, n-1\}$ that is one-to-one on S. The complexity measures of interest are storage space for h, evaluation time (which should be constant), and construction time. The talk gives an overview of several recent randomized constructions of minimal perfect hash functions, leading to space-efficient solutions that are fast in practice. A central issue is a method ("split-and-share") that makes it possible to assume that fully random (hash) functions are available.

1 Introduction

In this survey paper we discuss algorithmic techniques that are useful for the construction of minimal perfect hash functions. We focus on techniques for managing randomness.

We assume a set $U = \{0, 1\}^w$ (the "universe") of "keys" x is given. Assume that $S \subseteq U$ is a (given) set with cardinality $n = |S|$, and that $m \geq n$. A function $h: U \to [m]$ that is one-to-one on S is called a *perfect hash function* (for S). If in addition $n = m$ (the smallest possible value), h is called a *minimal perfect hash function* (MPHF).[1]

The MPHF problem for a given $S \subseteq U$ is to construct a data structure D_h that allows us to evaluate $h(x)$ for given $x \in U$, where h is a MPHF for S. The parameters of interest are the storage space for D_h and the evaluation time of h, which should be constant. Clearly, such a data structure D_h can be used to devise a (static) dictionary that for each key $x \in S$ stores x and some data item d_x in an array of size n, with constant retrieval time.

In the past decades, the MPHF problem has been studied thoroughly. For a detailed survey of the developments up to 1997 see the comprehensive study [9]. To put the results into perspective, one should notice the fundamental space lower bound of $n \log e + \log w - O(\log n)$ bits[2], valid as soon as $w \geq (2 + \varepsilon) \log n$, proved by Fredman and Komlós [18]. This bound is essentially tight: Mehlhorn [23, Sect. III.2.3, Thm. 8] gave a construction of a MPHF that takes $n \log e + \log w + O(\log n)$ bits of space (but has a vast evaluation time). In order not to have to worry about the influence of the size 2^w of U too much, unless

---

[1] $[m]$ denotes the set $\{0, \ldots, m-1\}$.
[2] All logarithms in this paper are to the base 2. Note that $\log e \approx 1.443 \ldots$

J. Hromkovič et al. (Eds.): SAGA 2007, LNCS 4665, pp. 2–17, 2007.
© Springer-Verlag Berlin Heidelberg 2007

noted otherwise, we will assume in the following that $n > w \geq (2 + \varepsilon) \log n$, and subsume the term $\log w$ in the space bounds in terms $O(\log n)$ and larger.

1.1 Space-Optimal, Time-Efficient Constructions

The (information-)theoretical background settled, the question is how close to the bound $n \log e + \log w$ one can get if one insists on constant evaluation time. In the seminal paper [19] Fredman, Komlós, and Szemerédi constructed a dictionary with constant lookup time, which can be used to obtain a MPHF data structure with constant evaluation time and space $O(n \log n)$ bits. Based on [19], Schmidt and Siegel [28] gave a construction for MPHF with constant evaluation time and space $O(n)$ bits (optimal up to a constant factor). Finally, Hagerup and Tholey [20] described a method that in expected linear time constructs a data structure D_h with $n + \log w + o(n + \log w)$ bits, for evaluating a MPHF h in constant time. This is space-optimal up to an additive term. It seems hard, though, to turn the last two constructions into data structures that are space efficient and practically time efficient at the same time for realistic values of n.

1.2 Practical Solutions

In a different line of development, methods for constructing MPHF were studied that emphasized the evaluation time and simple construction methods over optimality of space. Two different lines (a "graph/hypergraph-based approach" and a method called "hash-and-displace") in principle led to constructions of very simple structures that offered constant evaluation time and a space requirement that was dominated by a table of $\Theta(n)$ elements of $[n] = \{0, \ldots, n - 1\}$, which means $\Theta(n \log n)$ bits. Very recently, refinements of these methods were proposed that lead to a space requirement of $O(n \log \log n)$ bits (and constant evaluation time) [11,32]. Only in 2007, Botelho, Pagh, and Ziviani [5] managed to devise a construction for a MPHF that is simple and time-efficient, and gets by with $O(n)$ bits of storage space, with a constant factor that is only a small factor away from the information theory minimum $\log e \approx 1.44$. Crucial steps in this development will be described in some detail in the rest of this paper.

1.3 Randomness Assumptions

Given a universe U of keys, a *hash function* is just any function $h: U \to [m]$. Most constructions of MPHF involve several hash functions, which must behave randomly in some way or the other. There are two essentially different ways to approach the issue of the hash functions:

The "full randomness" assumption: One assumes that a sequence $h_0, h_1, \ldots$ of hash functions is available, so that evaluating $h_i(x)$ takes constant time, no storage space is needed for these functions, and such that $h_i(x), x \in S, i \geq 0$, are fully random values (uniform in $[m]$, independent). The analysis of several MPHF algorithms is based on this assumption (e. g., [8,22,7,4]).
Randomization: "Universal hashing" was introduced by Carter and Wegman [6] in 1979. One uses a whole set ("class") $\mathcal{H}$ of hash functions and chooses one such

function from $\mathcal{H}$ at random whenever necessary. Normally, some parameters of a function with a fixed structure are chosen at random. Storing the function means storing the parameters; the analysis is carried out on the basis of the probability space induced by the random choice of the function. Some classical MPFH algorithm use this approach (e. g., [28,25,20]).

Below, we will explain in detail how in the context of the MPHF problem one may quite easily work around the randomness issue by using very simple universal hash classes. To be concrete, we describe two such classes here. We identify $U = \{0,1\}^w$ with $[2^w]$.

Definition 1. *A set $\mathcal{H}$ of functions from U to $[m]$ is called 1-universal if for each pair of different $x, y \in U$ and for h chosen at random from $\mathcal{H}$ we have*

$$\Pr(h(x) = h(y)) \leq \frac{1}{m}.$$

There are many constructions of 1-universal classes. One is particularly simple (see [6]): Assume p is a prime number larger than 2^w, and $m \leq 2^w$. For $a, b \in [p]$ define $h_{a,b}(x) = ((ax + b) \bmod p) \bmod m$, and let $\mathcal{H}_m = \{h_{a,b} \mid a \in [p] - \{0\}, b \in [p]\}$. Choosing/storing a hash function from $\mathcal{H}_m$ amounts to choosing/storing the coefficients a and b (not much more than $2w$ bits).

Definition 2. *Let $k \geq 2$. A set $\mathcal{H}$ of functions from U to $[m]$ is called k-wise independent if for each sequence $(x_1, \ldots, x_k)$ of different elements of U and for h chosen at random from $\mathcal{H}$ we have that the values $h(x_1), \ldots, h(x_k)$ are fully random in $[m]^k$ and each value $h(x)$ is [approximately] uniformly distributed in $[m]$.*

The simplest way of obtaining a k-wise independent class is by using polynomials. Let $p > 2^w$ be a prime number as before, and let $m^{1+\varepsilon} \leq 2^w$ for some $\varepsilon > 0$. The set $\mathcal{H}_m^k$ of all functions of the form

$$h(x) = ((a_{k-1}x^{k-1} + \cdots + a_1 x + a_0) \bmod p) \bmod m, \quad a_{k-1}, \ldots, a_0 \in [p]$$

(polynomials over the field $\mathbf{Z}_p$ of degree smaller than k, projected into $[m]$), is k-wise independent. Choosing/storing a hash function amounts from this class amounts to choosing/storing the coefficients $(a_{k-1}, \ldots, a_0)$. For details see, e. g., [15,12]. The evaluation time is $\Theta(k)$. For more sophisticated hash function constructions see e. g. [29,14,30].

2   Split-and-Share for MPHFs

Let $S \subseteq U$ be fixed, $n = |S|$. For a hash function $h \colon S \to [m]$ and $i \in [m]$ let $S_i = \{x \in S \mid h(x) = i\}$, and let $n_i = |S_i|$. It is a common idea, used many times before in the context of perfect hashing constructions (e. g. in [19,20,10]), to construct separate and disjoint data structures for the "chunks" S_i.

The new twist is to "share randomness" among the chunks S_i, as follows. (The approach was sketched, for different applications, in [17,16].) In the static setting, with S given, this works as follows: Choose h, and calculate the sets $S_i = \{x \in S \mid h(x) = i\}$ and their sizes n_i, repeating if necessary until the sizes are suitable. Then devise one data structure that for each i provides one or several hash functions that behave fully randomly on S_i. Each S_i may own some component of this data structure but one essential part (usually a big table of random words) is used ("shared") by all S_i's.

We describe the approach in more detail. First, we "split", and make sure that none of the chunks is too large. The proof of the following lemma is standard.

Lemma 1. *If* $m \geq 2n^{2/3}$ *and* $h: U \to [m]$ *is chosen at random from a 4-universal class* $\mathcal{H} = \mathcal{H}_m^4$, *then* $\Pr(\max\{|S_i| \mid 0 \leq i < m\} > \sqrt{n}) \leq \frac{1}{4}$.

Proof. The probability that $|S_i| > \sqrt{n}$ is bounded by

$$\Pr\left(\binom{|S_i|}{4} \geq \binom{\sqrt{n}}{4}\right) \leq \frac{\mathrm{E}\left(\binom{|S_i|}{4}\right)}{\binom{\sqrt{n}}{4}} \leq \frac{\binom{n}{4}/(2n^{2/3})^4}{\binom{\sqrt{n}}{4}} < \frac{1}{8n^{2/3}},$$

for n large enough; hence $\Pr(\exists i: |S_i| \geq \sqrt{n}) \leq 2n^{2/3}/(8n^{2/3}) = \frac{1}{4}$.

Given S, we fix $m = 2n^{2/3}$ and repeatedly choose h from $\mathcal{H}_m^4$ until an h with $\max\{|S_i| \mid 0 \leq i < m\} \leq \sqrt{n}$ is found. We fix this function h and call it h^0 from here on; thus also the S_i and the n_i are fixed. With $a_i = \sum_{0 \leq j < i} n_j$ we can allocate indices in the interval $[a_i, a_{i+1} - 1]$ as possible hash values for keys in S_i.

Once we have found MPHFs h_i, one for each S_i, we may let

$$h(x) = a_i + h_i(x) \text{ for } i = h^0(x), \tag{1}$$

thus obtaining an MPHF for all of S. Below, we will describe several methods for building such a MPHF h_i. For this, it is most convenient to have at our disposal one or several hash functions that behave fully randomly (on each S_i separately). To make this concrete, let $K > 1$ be some constant, and let $L = K \log n$. We will argue that when considering S_i we may assume that we have a source of L fully random hash functions $h_1, \ldots, h_L$ from U to $\{0, 1\}^k$ for some k we may choose, which can be evalutated in (small) constant time. The data structure that provides the random elements used in these functions will be shared among the different h_i.

Let $\mathcal{H}_r$ denote an arbitrary 1-universal class of functions from U to $[r]$.

Lemma 2. *Let* $r = 2n^{3/4}$. *For an arbitrary given* $S' \subseteq U$ *with* $n' = |S'| \leq \sqrt{n}$ *we may in expected time* $O(|S'|)$ *find two hash functions* h_0, h_1 *from* $\mathcal{H}_r$ *such that for any two tables* $T_0[0..r-1]$ *and* $T_1[0..r-1]$, *each containing* r *random elements from* $\{0, 1\}^k$, *we have that* $h'(x) = T_0[h_0(x)] \oplus T_1[h_1(x)]$ *defines a function* $h' : U \to \{0, 1\}^k$ *that is fully random on* S'. ($\oplus$ *denotes bitwise XOR.*)

Proof. Assume h_0, h_1 are chosen at random from $\mathcal{H}_r$. We call a pair h_0, h_1 *good* if for each $x \in S'$ there is some $i \in \{0, 1\}$ such that $h_i(x) \neq h_i(y)$ for all $y \in S' - \{x\}$. For each $x \in S'$, the probability that $\exists y_0 \in S' - \{x\}: h_0(x) = h_0(y_0)$

and $\exists y_1 \in S' - \{x\}: h_1(x) = h_1(y_1)$ is smaller than $(\sqrt{n}/r)^2 \leq 1/(4\sqrt{n})$. This implies that the probability that (h_0, h_1) is not good is bounded by $\frac{1}{4}$. We keep choosing h_1, h_2 from $\mathcal{H}_r$ until a good pair is found — the expected number of trials is smaller than $\frac{4}{3}$. Checking one pair h_1, h_2 takes time $O(|S'|)$ when utilizing an auxiliary array of size r. Once a good pair h_1, h_2 has been fixed, for a key $x \in S'$ either table position $T_0[h_0(x)]$ or table position $T_1[h_1(x)]$ appears in the calculation of $h(x)$ but of no other key $y \in S'$. Since this entry is fully random, and because $\{0,1\}^k$ with $\oplus$ is a group, $h(x)$ is random and independent of the other hash values $h(y)$, $y \in S' - \{x\}$.

From here, we proceed as follows: For each i, $0 \leq i < m$, we choose hash functions h_0^i, h_1^i that are as required in Lemma 2 for $S' = S_i$. The descriptions of these $2m$ hash functions as well as the sizes n_i and the offsets a_i can be stored in (an array that takes) space $O(m) = O(n^{3/4})$ (words of length $O(w)$).

Now we describe the "shared" part of the data structure: Recall that $L = K \log n$. For each $j \in [L]$ we initialize arrays $T_{j,0}[0..r-1]$ and $T_{j,1}[0..r-1]$ with random words from $\{0,1\}^k$. We let

$$h_j^i(x) = T_{j,0}[h_0^i(x)] \oplus T_{j,1}[h_1^i(x)], \text{ for } x \in U, 0 \leq j < L, 0 \leq i < m.$$

Since h_0^i, h_1^i satisfy the condition in Lemma 2, for each fixed i we have that the values $h_{i,j}(x), x \in S_i, j \in [L]$, are fully random. The overall data structure takes up space $2n^{3/4} \cdot L$ words from $\{0,1\}^k$ plus $O(n^{2/3})$ words of size $\log |U|$, for the description of the h_0^i, h_1^i. We will see below that with high probability these hash functions will be sufficient for constructing a MPHF h_i for S_i, for all $i \in [m]$. If that construction is not successful, we start all over, with new random entries in the arrays $T_{j,0}$ and $T_{j,1}$.

From here on we *assume* that we have a fixed set S' of size $n' \leq \sqrt{n}$ and a supply of $L = K \log n$ fully random hash functions $h_0, \ldots, h_{L-1}$ with constant evaluation time and range $\{0,1\}^k$ (identified with $[2^k]$).

Goal: Build a MPHF for S' that has constant evaluation time and requires little storage space (beyond the functions $h_0, \ldots, h_{L-1}$). In the rest of the paper we discuss various strategies for achieving this.

3   Hash-and-Displace Approach

In this section, we discuss an approach to obtaining a MPHF by splitting S' into buckets, hashing the buckets into the common range $[n']$ and adjusting by offsets.

3.1   Pure Hash-and-Displace

Pagh [25] introduced the following approach for constructing a minimal perfect hash function for a set S': Choose hash functions $f: U \to [n']$ and $g: U \to [m']$. The set $[m'] \times [n']$ may be thought of as an array A with entry at (i, j) equal to 1 if $(f(x), g(x)) = 1$ for some $x \in S$, and 0 otherwise. Let $B_i = \{x \in S' \mid g(x) = i\}$,

$0 \leq i < m'$. We would like to see that f when restricted on B_i distributes the keys one-to-one into (the ith copy of) $[n']$. Technically, we check whether (f, g) is one-to-one on S' and whether

$$\sum_{\substack{0 \leq i < m' \\ |B_i| \geq 2}} |B_i|^2 < (1 - \delta)n'. \tag{2}$$

for some constant $\delta > 0$. Inequality (2) implies the "harmonic decay property" which is at the heart of the analysis of Pagh's algorithm:

$$s \cdot \sum_{\substack{0 \leq i < m' \\ |B_i| \geq s}} |B_i| < (1 - \delta)n', \text{ for all } s \geq 2. \tag{3}$$

If (2) is not satisfied, choose new (f, g) until (2) is satisfied. Pagh showed that once (2) is guaranteed a simple randomized scheme RFD ("random fit decreasing") in expected time $O(n')$ finds "displacements" $d_i \in [n']$, $0 \leq i < m'$, such that the function

$$h(x) = (f(x) + d_{g(x)}) \bmod n' \tag{4}$$

is (minimal) perfect for S'. Here, RFD works as follows: Sort the "rows" B_i by falling "weight" $|B_i|$. In this order, treat rows with $|B_i| \geq 2$ as follows: Repeat choosing d_i at random from $[n']$ until $\{(f(x) + d_i) \bmod n' \mid x \in B_i\}$ is disjoint from $\{(f(x) + d_{g(x)}) \bmod n' \mid x \in B_{i'}, B_{i'}$ already placed$\}$. Rows B_i with $|B_i| = 1$ are placed in one final deterministic round.

The question is how small m' may be chosen so that functions f and g as required can be found. Pagh based his construction on simple 1-universal classes and showed that to get by with f and g from such classes it is sufficient to have $m' > (2 + \varepsilon)n'$. Looking at (3) it is easy to see [13] that in place of (2) the following conditions are sufficient to make sure that RFD works: (f, g) is one-to-one on S' and

$$2 \cdot \sum_{\substack{0 \leq i < m' \\ |B_i| \geq 2}} |B_i| < (1 - \delta)n' \quad \text{and} \quad \sum_{\substack{0 \leq i < m' \\ |B_i| \geq 3}} |B_i|^2 < (1 - \delta)n'. \tag{5}$$

for some constant $\delta > 0$. With $\alpha = n'/m'$ one may show (with techniques explained in more detail in [13]) that asymptotically $(n', m' \to \infty)$

$$E\left(\sum_{\substack{0 \leq i < m' \\ |B_i| \geq 2}} |B_i| \right) \approx n' \cdot (1 - e^{-\alpha}) \quad \text{and} \quad E\left(\sum_{\substack{0 \leq i < m' \\ |B_i| \geq 3}} |B_i|^2 \right) \approx n' \cdot (\alpha + 1 - e^{-\alpha} - 2\alpha e^{-\alpha}). \tag{6}$$

Since for suitable $\varepsilon > 0$ and $\alpha < \ln 2/(1 + \varepsilon)$ we have $2(1 - e^{-\alpha}) < 1 - \delta$ and $\alpha + 1 - e^{-\alpha} - 2\alpha e^{-\alpha} < 0.6 < 1 - \delta$, this means that for n', m' large enough and $m' > 1.45(1 + \varepsilon)n' > (1 + \varepsilon)(\log e)n'$, inequalities (5) will be satisfied with high probability. Since the probability that (f, g) is not one-to-one on S' can be bounded by $\binom{n'}{2}/(n'm') < 1/(2 \cdot 1.45)$, a random pair (f, g) will be suitable with probability larger than $1/3$. If we try $(f, g) = (h_{2t} \bmod n', h_{2t+1} \bmod m'), t = 0, 1, \ldots, L/2 - 1$, for being one-to-one on S' and (5) being satisfied, the expected number of trials will be not larger than 3, the expected time will be $O(|S'|)$, and the probability we are not finished after testing $L/2$ pairs is at most $3^{-L/2} = 3^{-(K/2) \log n} = n^{-(K \log 3)/2}$. We can make this smaller than n^{-3} by choosing K large enough. Thus the probability that for some i no suitable (f, g) is found among $(h_{2t} \bmod n_i, h_{2t+1} \bmod m_i), t = 0, 1, \ldots, L/2 - 1$, is bounded by $m \cdot n^{-3} \leq n^{-2}$. (In this improbable case we choose new random entries for $T_{j,0}$ and $T_{j,1}$ and start all over.)

The overall data structure D_h for a hash-and-displace MPFH h for the whole set S consists of the following pieces:

- the splitting hash function h^0;
- h_0^i, h_1^i, for $0 \leq i < m$;
- values a_i (and b_i), for $0 \leq i < m$;
- arrays $T_{j,0}[0..r - 1]$ and $T_{j,1}[0..r - 1]$, for $j \in [L]$;
- an index $t \in [L/2]$ for the suitable pair $(h_{2t} \bmod n_i, h_{2t+1} \bmod m_i)$;
- $1.45(1 + \varepsilon)n_i$ offset values in $[n_i]$, for $0 \leq i < m$.

The overall space needed is $2m = 4n^{2/3}$ words of size $\log |U|$ and $mL = O(n^{2/3} \cdot \log n)$ words from $\{0, 1\}^k$ and $1.45(1 + \varepsilon)n$ offset values in $[\sqrt{n}]$ (about $0.78n \log n$ bits).

As remarked by P. Sanders [27], the space requirements may be lowered further asymptotically by increasing m to some larger power $n^{(t-1)/t}$, and increasing the degree of the splitting hash function h^0.

Remark: For the RFD algorithm to work, it is necessary that (f, g) are one-to-one, but condition (5) is only sufficient, not necessary. It is interesting to note that in (preliminary) experiments values of m' down to below $0.3n'$ still seem to work, so in a supervised situation where a MPHF is to be built (and one might resort to the pure, certified algorithm if not successful) it may save up to two thirds of the space if one tries to run RFD without checking (5).

3.2   Undo-One

Dietzfelbinger and Hagerup [13] modified Pagh's approach [25] as follows: Functions (f, g) were chosen to be one-to-one and satisfy the following conditions:

$$\sum_{\substack{0 \leq i < m' \\ |B_i| \geq 3}} |B_i|^2 < (1 - \delta)n' \quad \text{and} \quad |\{i \mid B_i \neq \emptyset\}| + |\{i \mid |B_i| = 1\}| \geq (1 + \delta)n'. \quad (7)$$

Once condition (7) is satisfied, a variant of Pagh's algorithm can be proved to find suitable offsets in expected constant time: For sets B_i with $|B_i| \geq 3$ run the RFD algorithm as before; for sets B_i of size 2 it is also checked whether they can be successfully be placed by moving up to one set that was placed before (for details see [13]).

It can be shown (using techniques from [13]) that for (f, g) fully random functions and $m' = (1 + \varepsilon)n'$, for an arbitrary fixed $\varepsilon > 0$, relation (7) holds with high probability, as long as n' is sufficiently large. This means that we may search for (f, g) just as in the previous section — only checking for (7) to hold. The final data structure will look the same as in the previous section. In contrast to [13], where for smaller ε polynomials of larger and larger degree are employed, the evaluation time for the hash functions described here does not depend on ε anymore. The overall space needed is $2m = 4n^{2/3}$ words of size $\log |U|$ and $n^{2/3}L = O(n^{2/3} \log n)$ words from $\{0, 1\}^k$ and $(1 + \varepsilon)n$ offset values in $[\sqrt{n}]$ (a little more than $0.5n \log n$ bits). Again, the constant factor in front of the $n \log n$ may be reduced at the expense of increasing m and the degree of independence of h^0.

4 Minimal Perfect Hashing by the Multifunction Paradigm

A different approach to constructing MPHF uses several hash functions. In that, it resembles the approach taken in the area of Bloom filters (see [3], and e. g. [7]).

4.1 The Hypergraph Approach

Czech *et al.* [8] and Majewski *et al.* [22] introduced the following approach to constructing a MPHF. To each key x associate a sequence $(h_1(x), \ldots, h_d(x))$ of *distinct* hash values in some range $[m]$. The structure consisting of $V = [m]$ and the system of (labeled) sets $e_x = \{h_1(x), \ldots, h_d(x)\}$, $x \in S'$, may be regarded as a hypergraph $G(S', h_1, \ldots, h_d)$ of order (edge size) d. If the elements of S' can be arranged in a sequence $(x_1, \ldots, x_{n'})$ such that

$$e_{x_j} - \bigcup_{s<j} e_{x_s} \neq \emptyset \ , \text{ for } j = 1, \ldots, n' \tag{8}$$

then we say that $G(S', h_1, \ldots, h_d)$ is *acyclic*. It is useful to consider the vertex-edge incidence matrix A_G of $G(S', h_1, \ldots, h_d)$. It has n' rows, labeled with $x_1, \ldots, x_{n'}$, where position ℓ in row j is 1 if $\ell \in \{h_1(x_j), \ldots, h_d(x_j)\}$, and is 0 otherwise. Condition (8) entails that in this matrix in row j there is a 1-entry in some position ℓ_j so that column ℓ_j has only 0s above row j. Thus the matrix A_G can be transformed into echelon form by exchanging columns. This immediately implies the following.

Lemma 3. *If (8) holds, then for each vector $(b_1, \ldots, b_{n'}) \in [n']$ we may (even in linear time) find a set of values $g(i) \in [n']$, $i \in [m']$, such that*

$$(g(h_1(x_j)) + \cdots + g(h_d(x_j))) \bmod m' = b_j \ , \text{ for } 1 \leq j \leq n'.$$

It can be arranged that $g(i) = 0$ for $i \notin \{\ell_j \mid 1 \leq j \leq n'\}$. If we choose $(b_1, \ldots, b_{n'})$ as a permutation of $(0, 1, \ldots, n' - 1)$, we obtain a MPHF for S'.

Remark 1. In [7] the approach of the acyclic hypergraph was re-discovered and utilized in a similar way as in [22] to implement an arbitrary function $f: S' \to \{0, 1\}^q$, even including a mechanism to detect (with some probability $1 - \varepsilon$)

if a key $x \notin S'$ is presented to the data structure. (This problem was called "retrieval" in [10].) In [7], a naive analysis of the acyclicity property was used, leading to an estimate $O(n')$ of the space requirements that is much larger than the space bounds from [22], as discussed below. However, the bounds from [22] do apply also in this context.

In [8] and [22] it was assumed that fully random hash functions are available. This gap in the analysis vanishes if one employs the "split-and-share" trick.

A minor question that remains is how one may find a mapping $x \mapsto (h_1(x), \ldots, h_d(x))$ that attains vectors of d different elements as values, each one with the same probability, if only given fully random values $(h_1^0(x), \ldots, h_d^0(x))$ in $\{0, 1\}^k$. There are several approaches to this problem, a solution due to Floyd being discussed in [2]. (A workaround used in some papers (e.g. [5]) is to let $h_1, \ldots, h_d$ have disjoint ranges of size m'/d each, the slight disadvantage being that results from the random graphs literature do not apply directly to this situation of "d-partite hypergraphs".)

Again, with the "splitting" approach at the basis, we do not have to worry about space, and can even simplify Floyd's method, utilizing an idea usually employed to construct (full) random permutations (see [21, Algorithm P]). Use an auxiliary array $R[0..\sqrt{n}-1]$, initialized so that $R[i] = i$ for all i (this property is restored after each use). We assume that a fully random sequence $h_1^0, \ldots, h_d^0$ of hash functions with range $[2^k]$ is available, $k \geq 2 \log n$.

Algorithm. *Hyperedge*
Input: x, n'.
Output: $(h_1(x), \ldots, h_d(x))$ ($*$ distinct values $*$)
for $\ell = 1$ **to** d **do**
    $j_\ell \leftarrow h_\ell^0(x) \bmod (n' - \ell + 1)$;
    exchange $R[j_\ell]$ and $R[n' - \ell]$;
$(z_1, \ldots, z_d) \leftarrow (R[n'-1], \ldots, R[n'-d])$;
for $\ell = 1$ **to** d **do**
    $R[j_\ell] \leftarrow j_\ell$; $R[n'-\ell] \leftarrow n'-\ell$;
return $(z_1, \ldots, z_d)$.

It is not hard to check that each d-tuple $(z_1, \ldots, z_d)$ in $[n']$ that consists of d distinct values (up to negligibly small rounding errors) has the same probability to be returned as $(h_1(x), \ldots, h_d(x))$. Once the edges e_x, $x \in S'$, have been calculated, one may easily in linear time calculate an ordering $(x_1, \ldots, x_{n'})$ that satisfies (8), if such an ordering exists. (For details see [22].) If no such ordering exists (the hypergraph $([m], \{e_x\}_{x \in S'})$ is "cyclic"), we repeat with a new set $h_1^0, \ldots, h_d^0$ of fully random hash functions. (If this approach is implemented in the context of the "split-and-share" approach, for each trial a new segment of d of the fully random functions $h_0, \ldots, h_{L-1}$ are used.)

In [22] it is discussed in detail what the probability for acyclicity is for various d and quotients $c = n'/m'$ (assuming the asymptotic case with $n', m' \to \infty$). For $d = 2$ we must have $c > 2$ and get an acyclicity probability of $e^{1/c}\sqrt{(c-2)/c} > 0$.

For $d = 3, 4, 5$ one gets threshold values

$$c_3 \approx 1.222, c_4 \approx 1.295, c_5 \approx 1.425,$$

meaning that if $n'/m' \geq c > c_d$ then the probability that the hypergraph is acylic is high (approaching 1 as n', m' grow). Larger values of d have worse threshold values. The most attractive choice for d obviously is $d = 3$, where a choice of $m' = 1.23n'$ leads to a good chance for hitting an acyclic hypergraph.

Thus, a data structure for a mapping $U \ni x \mapsto e_x = \{h_1(x), \ldots, h_d(x)\}$ so that $([m'], \{e_x\}_{x \in S'})$ forms an acyclic hypergraph can be constructed in expected time $O(|n'|)$. We have already seen how such a structure can be used to get a MPHF for S' that in essence consists of a table of m' numbers from $[\sqrt{n}]$. If one uses this construction for each set S_i separately, sharing the random entries in the arrays $T_{j,0}, T_{j,1}$ as before, one obtains a data structure D_h for a MPFH h for S that has the following components:

- the splitting hash function h^0;
- h_0^i, h_1^i, for $0 \leq i < m$;
- values a_i (and b_i), $0 \leq i < m$;
- arrays $T_{j,0}[0..r-1]$ and $T_{j,1}[0..r-1]$, $j \in [L]$;
- an index $t \in [L/2]$ for the suitable triple
  $(h_{3t} \bmod n_i, h_{3t+1} \bmod m_i, h_{3t+2} \bmod m_i)$;
- $1.23n_i$ g_i-values in $[n_i]$, for $0 \leq i < m$.

The overall space needed is $2m = 4n^{2/3}$ words of size $\log|U|$ and mL words from $\{0,1\}^k$ and $1.23n$ offset values in $[\sqrt{n}]$ (about $0.62n \log n$ bits).

5   Below the Graph Thresholds

Using the approach of Majewski *et al.* [22] one may not get below the space bound $1.23n'$ given by the requirement that the random hypergraphs be acyclic. The Undo-Une construction from [13] achieves space $(1+\varepsilon)n'$, but seemingly not less. However, in [4] and in [31] methods for constructing MPHF are described that have the potential to get below the threshold of n' words. Botelho *et al.* [4] as well as Weidling [31] independently propose using the hypergraph approach with $d = 2$, in which case the hypergraph $G(S', h_1, h_2)$ turns into a standard graph. The central change is to give up the requirement that this graph be acyclic. Rather, these authors propose studying the 2-*core* $J_2 \subseteq [m']$ of $G(S', h_1, h_2)$, which is the largest subgraph all of whose nodes having degree 2 or larger.

From graph theory it is well known that the 2-core of a graph G can be found in linear time by a simple "peeling" process. This process iterates cutting off nodes of degree 1 ("leaves") from G; the remaining graph with nodes of degree at least 2 is the 2-core. If one assumes that the 2-core J_2 of $G(S', h_1, h_2)$ has been determined and that values $g(i)$, $i \in J_2$, have been calculated such that

$$\text{the mapping } x \mapsto (g(h_1(x)) + g(h_2(x))) \bmod n' \tag{9}$$

is one-to-one on the set $\{x \in S' \mid h_1(x), h_2(x) \in J_2\}$, then it is very easy to calculate values $g(i)$ for $i \in [m'] - J_2$ such that $x \mapsto (g(h_1(x)) + g(h_2(x))) \bmod n'$ is one-to-one on S'. (See Section 5.2.) We are left with the problem of finding a suitable g-labeling of the nodes in J_2. Here, the methods of [4] and [31] differ.

5.1   A Partly Heuristic Approach

Botelho *et al.* [4] propose a greedy strategy for determining the g-values inside J_2. This strategy assigns g-values to nodes in the order $i_1, \ldots, i_{|J_2|}$ of a breadth-first-search in the 2-core, where each value $g(i_t)$ is chosen so that it is bigger than $g(i_1), \ldots, g(i_{t-1})$ but minimal so as not to get into conflict with (9). The authors of [4] conjecture (Conjecture 1, [4, p. 496]) that if

$$\text{the 2-core of } G(S', h_1, h_2) \text{ has } \leq n'/2 \text{ edges,} \tag{10}$$

then this greedy strategy succeeds in the sense that the largest g-value assigned is not larger than $n' - 1$. For this conjecture, experimental evidence is provided.

It remains to estimate the edge density n'/m' we can afford so that with high probability the 2-core has no more than $n'/2$ edges. Referring to results on the structure of 2-cores of random graphs, in particular to [26], in [4] the following rule is provided (valid for $n', m' \to \infty$): The number of edges in the 2-core is

$$(1 + o(1))(1 - T/d)^2 n',$$

where $d = 2n'/m'$ is the average degree of $G(S', h_1, h_2)$ and T is the unique solution in $(0, 1)$ of the equation $Te^{-T} = de^{-d}$. A simple numeric computation shows that the threshold value for d is approximately 1.736. This means that (for n', m' large) we can afford $d \approx 1.73$, or $m' \geq 1.152n'$, and may expect to have no more than $n'/2$ edges in the 2-core.

In [4] experimental evidence is given that the algorithm works well for this choice of m'. The authors further report that in experiments a variant of their algorithm (not insisting that the values $g(x_t)$ increase with t increasing) makes it possible to further decrease m', to some value $m'/n' \approx 0.93$, but not further.

5.2   An Analyzed Approach

We turn to Weidling's [31] analysis of the strategy based on the 2-core of $G(S', h_1, h_2)$. The lowest edge density we consider in the analysis is never larger than 1.1, meaning that always $m' > 0.9n'$. In this case the probability that $G(S', h_1, h_2)$ has an empty 3-core (i.e., $G(S', h_1, h_2)$ does not have a nonempty subgraph with minimum degree 3) is overwhelming. Further, with high probability the maximum degree of nodes in $G(S', h_1, h_2)$ is $O(\log(n'))$. Finally, with positive probability $G(S', h_1, h_2)$ does not have double edges. For simplicity from here on we assume that $G(S', h_1, h_2)$ satisfies these properties (otherwise this will turn out at some time of the execution of the algorithm, in which case we choose new hash functions h_1, h_2 for S'). The following assumption is crucial for the analysis of the first algorithm (cf. (10)):

$$\text{the 2-core of } G(S', h_1, h_2) \text{ has } \leq (\tfrac{1}{2} - \varepsilon)n' \text{ edges.} \tag{11}$$

We "peel" $G(S', h_1, h_2)$, as follows: Let $G_1 = G(S', h_1, h_2)$. We disregard nodes of degree 0.

- Round $t = 1, \ldots, \ell_1$: We choose a node j_t of degree 1 in G_t and obtain G_{t+1} by removing j_t and the (unique) incident edge $\{h_1(x_t), h_2(x_t)\}$. — What remains is a graph G_{ℓ_1} with minimum degree 2, the 2-core.

Round $t = \ell_1 + 1, \ldots, \ell_2$: If all nodes in G_t have degree 2 or larger, we choose a node j_t of degree 2 and obtain G_{t+1} by removing j_t and its two incident edges $\{h_1(x_{t,1}), h_2(x_{t,1})\}$, $\{h_1(x_{t,2}), h_2(x_{t,2})\}$ from G_t. Otherwise we choose a node j_t of degree 1, and obtain G_t by removing it and the (unique) incident edge $\{h_1(x_t), h_2(x_t)\}$. This is continued until an empty graph results.

Now the g-values are assigned. Preliminarily, assign g-value 0 to all nodes $j \in [m']$. Let $H = \emptyset$ (the already assigned hash values). We proceed in the reverse order of the peeling process.

Round $t = \ell_2, \ldots, \ell_1 + 1$:

Case 1: Node j_t has degree 2 in G_t, with two incident edges $\{h_1(x_{t,1}), h_2(x_{t,1})\}$, $\{h_1(x_{t,2}), h_2(x_{t,2})\}$. Assume $j_t = h_1(x_{t,1}) = h_1(x_{t,2})$. (The other cases are treated analogously.) Let $j' = h_2(x_{t,1})$ and $j'' = h_2(x_{t,2})$. What are legal values for $g(j_t)$ so that we do not get stuck on our way to constructing a MPHF? We must have

(i) $(g(j_t) + g(j')) \bmod n'$, $(g(j_t) + g(j'')) \bmod n'$ are *different* and not in H, and

(ii) $g(j_t) \notin \{g(j_s) \mid s > t$ and j_s has distance 2 to $j_t\}$.

That condition (i) is necessary (and sufficient for carrying out step t) is obvious. Condition (ii) makes sure that in a later step t' it will not happen that a common neighbor of j_t and $j_{t'}$ cannot be labeled because $g(j_t) = g(j_{t'})$. Since by assumption (11) we have $|H| \leq \frac{1}{2}(1 - \varepsilon)n'$, and since graph $G(S', h_1, h_2)$ has maximum degree $O(\log(n'))$, conditions (i) and (ii) exclude at most $(1 - \varepsilon)n' - O(\log(n'))$ values for $g(j_t)$. Since there are n' values to choose from, we may try values from $[n']$ at random until a suitable value for $g(j_t)$ is found, and will succeed after an expected number of $1/\varepsilon$ rounds. One still has to prove the simple fact that the expected number of nodes at distance 1 and 2, averaged over all nodes, is $O(1)$, to conclude that the overall construction time is expected $O(n)$.

Case 2: Node j_t has degree 1 in G_t. Let j' be its unique neighbor in G_t. In this case the new value $g(j_t)$ just has to satisfy (i)' $(g(j_t) + g(j')) \bmod n' \notin H$ and condition (ii); again after an expected constant number of random trials we will find a suitable value $g(j_t)$.

- Round $t = \ell_1, \ldots, 1$:

We know that node j_t has degree 1 in G_t. Let j' be its unique neighbor. Deterministically choose $g(j_t) = (i + n' - g(j')) \bmod n'$ for one (the next) element $i \notin H$.

This algorithm finishes in expected linear time, if (11) is satisfied. On the same grounds from random graph theory as noted in Section 5.1 one sees that for this the condition $m' \geq 1.152(1 + \delta)n'$ for some $\delta > 0$ is sufficient. (In [31] a direct estimate of the number of edges in the 2-core is provided, leading to the same result.)

Looking from the point of view of space efficiency, the threshold $m' \geq 1.152$ $(1 + \delta)n'$ is not yet satisfying since the construction from [13] achieves a similar result with space $m' = (1 + \delta)n'$.

Weidling [31] investigated the combination of the graph-based construction with the "Undo-One" strategy from Section 3.2. He proved that $m' \geq 0.9353(1 + \delta)n'$ is sufficient to guarantee that an adapted version of this strategy succeeds in building a graph-based MPHF. For the full data structure in the context of the split-and-share approach this would lead to the same space requirements as in Section 4.1, replacing the term "$1.23n$ offset values in $[\sqrt{n}]$ (about $0.62n \log n$ bits)" by "$0.94n$ offset values in $[\sqrt{n}]$ (about $0.47n \log n$ bits)".

6   Below $n \log n$

Using methods different from those described in this paper, based on the approach of [20], Woelfel [32] provided a MPHF construction that had more practical evaluation times than the purely theoretical constructions but gets by with space $O(n \log \log n)$. A similar result was reported in [11]. This construction is based on the hash-and-displace approach with the random-fit-decreasing algorithm, see 3.1. However, for each bucket B_i of size 2 or larger a new sequence of fully random hash functions is employed (instead of one fixed f for all buckets). Only the index of the successful hash function has to be stored, which will be a number of size $O(\log n)$, hence of $\log \log n + O(1)$ bits. The buckets B_i of size 1 cause a new subtle problem. To allocate these buckets with a hash function out of a pool of $O(\log n)$ many, one has to construct a perfect matching in the graph induced by the x's in such buckets and the respective hash values $h_0, \ldots, h_{L-1}$, for $L = \Theta(\log n)$. For this, methods for finding matchings in sparse random graphs are employed ([1,24]). The construction time rises to $O(n(\log n)^2)$.

7   An Almost Optimal Solution

Very recently, Botelho, Pagh, and Ziviani ([5], WADS'07) described a method to obtain a MPFH with description size $O(n)$, a constant factor away from the optimum. We give a brief account of their approach. For the theoretical analysis, they appeal to the "split-and-share" approach just as we did before, so we may assume that we have to achieve the goal formulated at the end of Section 1.3: find a MPHF for S', $n' = |S'| \leq \sqrt{n}$, assuming a pool of $K \log n$ fully random functions. Botelho $et\ al.$ set out from the hypergraph setting of Section 4.1, with hypergraphs of order d. They use the fact known from random graph theory that for each $d \geq 2$ there is a constant c_d such that if $m'/n' \geq c > c_d$ then the hypergraph $G(S', h_1, \ldots, h_d)$ is acyclic (with positive probability for $d = 2$ and with high probability as $n', m' \to \infty$ for $d \geq 3$). They calculate the corresponding order $(x_1, \ldots, x_{n'})$ of the elements of S' such that (8) is satisfied. The crucial observation now is that the mapping

$$x_j \mapsto \text{some element } \ell_j \text{ of } e_{x_j} - \bigcup_{s<j} e_{x_s}$$

from S' to $[m']$ is one-to-one. Thus, for each j we may choose some $\ell_j \in [d]$ (namely, an arbitrary $\ell_j \in e_{x_j} - \bigcup_{s<j} e_{x_s}$) such that the mapping

$$h': U \ni x \mapsto h_{\ell_j+1}(x) \in [m'] \tag{12}$$

is one-to-one on S'. Thus this mapping already represents a hash function that is perfect on S', with a range that is a little larger than we would like it to be.

A central idea of [5] now is to provide a data structure for calculating h'. (As mentioned in Remark 1, this idea was already used in very much the same way by Chazelle *et al.* [7] in the context of data structures that represent "half-dynamic" mappings from S' to some range X, there called "Bloomier filters", but without an attempt to make m' as small as possible.) This is done as follows: We use the construction indicated in Lemma 3, but not for $[n']$ with modular addition, but for $[d]$: We may find, in linear time, values $g(i) \in [d]$, $i \in [m']$, such that

$$(g(h_1(x_j)) + \cdots + g(h_d(x_j))) \bmod d = \ell_j \text{ , for } 1 \le j \le n'.$$

Moreover, we arrange that $g(i) = 0$ for $i \notin \{\ell_j \mid 1 \le j \le n'\}$. These values $g(i)$, when stored in a table with m' entries from $[d]$, form a data structure that makes it possible to calculate $h'(x)$, $x \in U$, from (12) as follows:

$$h'(x) = h_{1+(g(h_1(x))+\cdots+g(h_d(x))) \bmod d}(x). \tag{13}$$

Storing the values $g(j), j \in [m']$, takes about m' blocks of $\lceil \log d \rceil$ bits, or, by coding s numbers into one block of length $\lceil \log(d^s) \rceil$, about m'/s blocks of $\lceil \log(d^s) \rceil$ bits. For a concrete figure, let $d = 3$, in which case one may use $m' = 1.23n'$, and $s = 5$. Then $\lceil \log(d^s) \rceil = \lceil \log(243) \rceil = 8$, so a block is a byte, and one needs $1.23n$ bytes or approximately $1.97n'$ bits. Using this approach for constructing a perfect hash function for S one obtains a data structure that uses no more than $2n + O(n^{2/3} \log n)$ bits, and has an extremely simple structure.

The hash function h' described so far does not have minimal range $[n']$. In [5] the following approach is proposed. In the table for the g-values positions $j \in [m'] - \{\ell_j \mid 1 \le j \le n'\}$ are filled with the entry "d" — indicating that the index does not belong to the set $\{\ell_j \mid 1 \le j \le n'\}$, but having no effect on the arithmetic modulo d. Thus, for $d = 3$ one has four possible g-values, requiring 2 bits per entry, resulting in a little bit more than $2.46n'$ bits.

The MPFH h we aim at is defined as

$$h(x) = |\{\ell_j \mid 1 \le j \le n', \ell_j < h'(x)\}|. \tag{14}$$

A moment's thought reveals that this function h takes on values in $[n']$ and is one-to-one on S', because the $h'(x)$-values for the elements $x \in S'$ are just the n' elements in $\{\ell_j \mid 1 \le j \le n'\}$. There are several ways of calculating $h(x)$ using the table of g-values and some auxiliary data structure — for details see [5].

The authors of [5] report on experiments that indicate that their approach leads to data structures that are space-efficient as described by the theory as well as very time-efficient. (The implementations use standard universal hash classes, which turn out to be sufficient so that it is not necessary to employ the "split-and-share" trick.)

8 Conclusion

The study of data structures for the MPHF problem has taken some interesting steps in the past few years. From the theoretical side, it has been understood how the full randomness assumption may be justified without resorting to constructions with large evaluation times, and — building on earlier work that have demonstrated the crucial role played by hypergraph structures — a practically useful construction has been found that approaches the theoretically optimal space bound of $n \log e$ bits up to a small constant factor. A natural question left open is whether one can get even closer to the space lower bound, while retaining practicability. Also, it would be interesting to see whether provable randomness properties remain in the graph and hypergraph structures discussed here if one does not assume full randomness but only k-wise independence for some k.

References

1. Bast, H., Mehlhorn, K., Schäfer, G., Tamaki, H.: Matching algorithms are fast in sparse random graphs. Theory Comput. Syst. 39(1), 3–14 (2006)
2. Bentley, J.: Programming pearls: a sample of brilliance. J. Assoc. Comput. Mach. 30(9), 754–757 (1987)
3. Bloom, B.H.: Space/time trade-offs in hash coding with allowable errors. Commun. ACM 13(7), 422–426 (1970)
4. Botelho, F.C., Kohayakawa, Y., Ziviani, N.: A practical minimal perfect hashing method. In: Nikoletseas, S.E. (ed.) WEA 2005. LNCS, vol. 3503, pp. 488–500. Springer, Heidelberg (2005)
5. Botelho, F.C., Pagh, R., Ziviani, N.: Simple and space-efficient minimal perfect hash functions. In: Proc. 10th Workshop on Algorithms and Data Structures (WADS 2007). LNCS, vol. 4619, Springer, Heidelberg (to appear 2007)
6. Carter, L., Wegman, M.N.: Universal classes of hash functions. J. Comput. Syst. Sci. 18(2), 143–154 (1979)
7. Chazelle, B., Kilian, J., Rubinfeld, R., Tal, A.: The Bloomier filter: an efficient data structure for static support lookup tables. In: Proc. 15th ACM-SIAM SODA 2004, pp. 30–39 (2004)
8. Czech, Z.J., Havas, G., Majewski, B.S.: An optimal algorithm for generating minimal perfect hash functions. Inform. Proc. Lett. 43(5), 257–264 (1992)
9. Czech, Z.J., Havas, G., Majewski, B.S.: Perfect Hashing (Fundamental Study). Theor. Comput. Sci. 182(1–2), 1–143 (1997)
10. Demaine, E.D., Meyer auf der Heide, F., Pagh, R., Patrascu, M.: De dictionariis dynamicis pauco spatio utentibus (lat. On dynamic dictionaries using little space). In: Correa, J.R., Hevia, A., Kiwi, M.A. (eds.) LATIN 2006. LNCS, vol. 3887, pp. 349–361. Springer, Heidelberg (2006)
11. Dietzel, L.: Speicherplatzeffiziente perfekte Hashfunktionen, Diplomarbeit, Technische Universität Ilmenau, Fakultät IA (in German) (2005)
12. Dietzfelbinger, M., Gil, J., Matias, Y., Pippenger, N.: Polynomial hash functions are reliable. In: Kuich, W. (ed.) Automata, Languages and Programming. LNCS, vol. 623, pp. 235–246. Springer, Heidelberg (1992)
13. Dietzfelbinger, M., Hagerup, T.: Simple minimal perfect hashing in less space. In: Meyer auf der Heide, F. (ed.) ESA 2001. LNCS, vol. 2161, pp. 109–120. Springer, Heidelberg (2001)

14. Dietzfelbinger, M., Meyer auf der Heide, F.: Dynamic hashing in real time. In: Paterson, M.S. (ed.) Automata, Languages and Programming. LNCS, vol. 443, pp. 6–19. Springer, Heidelberg (1990)
15. Dietzfelbinger, M., Karlin, A., Mehlhorn, K., Meyer auf der Heide, F., Rohnert, H., Tarjan, R.: Dynamic perfect hashing: Upper and lower bounds. SIAM J. Comput. 23(4), 738–761 (1994)
16. Dietzfelbinger, M., Weidling, C.: Balanced allocation and dictionaries with tightly packed constant size bins. Theoret. Comput. Sci. 380(1–2), 47–68 (2007)
17. Fotakis, D., Pagh, R., Sanders, P., Spirakis, P.G.: Space efficient hash tables with worst case constant access time. Theory Comput. Syst. 38(2), 229–248 (2005)
18. Fredman, M.L., Komlós, J.: On the size of separating systems and families of perfect hash functions. SIAM J. Alg. Disc. Meth. 5(1), 61–68 (1984)
19. Fredman, M.L., Komlós, J., Szemerédi, E.: Storing a sparse table with 0(1) worst case access time. J. Assoc. Comput. Mach. 31(3), 538–544 (1984)
20. Hagerup, T., Tholey, T.: Efficient minimal perfect hashing in nearly minimal space, in: Proc. 18th STACS 2001, Springer LNCS 2010. In: Ferreira, A., Reichel, H. (eds.) STACS 2001. LNCS, vol. 2010, pp. 317–326. Springer, Heidelberg (2001)
21. Knuth, D.E.: The Art of Computer Programming, vol. 2: Seminumerical Algorithms, 2nd edn. Addison-Wesley, Reading (1981)
22. Majewski, B.S., Wormald, N.C., Havas, G., Czech, Z.J.: A family of perfect hashing methods. Computer J. 39(6), 547–554 (1996)
23. Mehlhorn, K.: Data Structures and Algorithms, Vol. 1: Sorting and Searching. Springer, Berlin (1984)
24. Motwani, R.: Average-case analysis of algorithms for matchings and related problems. J. Assoc. Comput. Mach. 41(6), 1329–1356 (1994)
25. Pagh, R.: Hash and displace: Efficient evaluation of minimal perfect hash functions. In: Dehne, F., Gupta, A., Sack, J.-R., Tamassia, R. (eds.) WADS 1999. LNCS, vol. 1663, pp. 49–54. Springer, Heidelberg (1999)
26. Pittel, B., Wormald, N.C.: Counting connected graphs inside-out. J. Comb. Theory, Ser. B 93(2), 127–172 (2005)
27. Sanders, P.: personal communication
28. Schmidt, J.P., Siegel, A.: The spatial complexity of oblivious k-probe hash functions. SIAM J. Comput. 19(5), 775–786 (1990)
29. Siegel, A.: On universal classes of extremely random constant-time hash functions. SIAM J. Comput. 33(3), 505–543 (2004)
30. Thorup, M.: Even strongly universal hashing is pretty fast. In: Proc. 11th ACM-SIAM SODA 2000, pp. 496–497 (2000)
31. Weidling, C.: Platzeffiziente Hashverfahren mit garantierter konstanter Zugriffszeit, Dissertation. Technische Universität Ilmenau (in German) Electronic version. (2004), http://www.db-thueringen.de/servlets/DocumentServlet?id=2431
32. Woelfel, P.: Maintaining external memory efficient hash tables. In: Díaz, J., Jansen, K., Rolim, J.D.P., Zwick, U. (eds.) APPROX 2006 and RANDOM 2006. LNCS, vol. 4110, pp. 508–519. Springer, Heidelberg (2006)

Hamming, Permutations and Automata*

Rūsiņš Freivalds

Institute of Mathematics and Computer Science, University of Latvia,
Raiņa bulvāris 29, Rīga, Latvia

Abstract. Quantum finite automata with mixed states are proved to be super-exponentially more concise rather than quantum finite automata with pure states. It was proved earlier by A.Ambainis and R.Freivalds that quantum finite automata with pure states can have exponentially smaller number of states than deterministic finite automata recognizing the same language. There was a never published "folk theorem" proving that quantum finite automata with mixed states are no more than super-exponentially more concise than deterministic finite automata. It was not known whether the super-exponential advantage of quantum automata is really achievable.

We prove that there is an infinite sequence of distinct integers n such that there are languages L_n such that there are quantum finite automata with mixed states with $5n$ states recognizing the language L_n with probability $\frac{3}{4}$ while any deterministic finite automaton recognizing L_n needs to have at least $e^{O(n \ln n)}$ states.

Unfortunately, the alphabet for these languages grows with n. In order to prove a similar result for languages in a fixed alphabet we consider a counterpart of Hamming codes for permutations of finite sets, i.e. sets of permutations such that any two distinct permutations in the set have Hamming distance at least d. The difficulty arises from the fact that in the traditional Hamming codes for binary strings positions in the string are independent while positions in a permutation are not independent. For instance, any two permutations of the same set either coinside or their Hamming distance is at least 2. The main combinatorial problem still remains open.

1 Introduction

A.Ambainis and R.Freivalds proved in [4] that for recognition of some languages the quantum finite automata can have smaller number of the states than deterministic ones, and this difference can even be exponential. The proof contained a slight non-constructiveness, and the exponent was not shown explicitly. For probabilistic finite automata exponentiality of such a distinction was not yet proved. The best (smaller) gap was proved by Ambainis [2]. The languages recognized by automata in [4] were presented explicitly but the exponent was not. In a very recent paper by R.Freivalds [10] the non-constructiveness is modified,

---

* Research supported by Grant No.05.1528 from the Latvian Council of Science.

J. Hromkovič et al. (Eds.): SAGA 2007, LNCS 4665, pp. 18–29, 2007.
© Springer-Verlag Berlin Heidelberg 2007

and an explicit (and seemingly much better) exponent is obtained at the expense of having only non-constructive description of the languages used. Moreover, the best estimate proved in this paper is proved under assumption of the well-known Artin's Conjecture (1927) in Number Theory. [10] contains also a theorem that does not depend on any open conjectures but the estimate is worse, and the description of the languages used is even less constructive. This seems to be the first result in finite automata depending on open conjectures in Number Theory.

The following two theorems are proved in [10]:

Theorem 1. *Assume Artin's Conjecture. There exists an infinite sequence of regular languages $L_1, L_2, L_3, \ldots$ in a 2-letter alphabet and an infinite sequence of positive integers $z(1), z(2), z(3), \ldots$ such that for arbitrary j:*

1. *there is a probabilistic reversible automaton with $(z(j)$ states recognizing L_j with the probability $\frac{19}{36}$,*
2. *any deterministic finite automaton recognizing L_j has at least $(2^{1/4})^{z(j)} = (1.1892071115\ldots)^{z(j)}$ states,*

Theorem 2. *There exists an infinite sequence of regular languages $L_1, L_2, L_3, \ldots$ in a 2-letter alphabet and an infinite sequence of positive integers $z(1), z(2), z(3), \ldots$ such that for arbitrary j:*

1. *there is a probabilistic reversible automaton with $z(j)$ states recognizing L_j with the probability $\frac{68}{135}$,*
2. *any deterministic finite automaton recognizing L_j has at least $(7^{\frac{1}{14}})^{z(j)} = (1.1149116725\ldots)^{z(j)}$ states,*

The two theorems above are formulated in [10] as assertions about reversible probabilistic automata. For probabilistic automata (reversible or not) it was unknown before the paper [10] whether the gap between the size of probabilistic and deterministic automata can be exponential. It is easy to re-write the proofs in order to prove counterparts of Theorems 1 and 2 for quantum finite automata with pure states. The aim of this paper is to prove a counterpart of these theorems for quantum finite automata with mixed states.

Quantum algorithms with mixed states were first considered by D.Aharonov, A.Kitaev, N.Nisan [1]. More detailed description of quantum finite automata with mixed states can be found in A.Ambainis, M.Beaudry, M.Golovkins, A.Ķikusts, M.Mercer, D.Thrien [3].

The automaton is defined by the initial density matrix ρ_0. Every symbol a_i in the input alphabet is associated with a unitary matrix A_i. When the automaton reads the symbol a_i the current density matrix ρ is transformed into $A_i^* \rho A_i$. When the reading of the input word is finished and the end-marker $ is read, the current density matrix ρ is transformed into $A_{end}^* \rho A_{end}$ and separate measurements of all states are performed. After that the probabilities of all the accepting states are totalled, and the probabilities of all the rejecting states are totalled.

Like quantum finite automata with pure states described by A.Kondacs and J.Watrous [12] we allow measurement of the accepting states and rejecting states after every step of the computation.

The main result in our paper is:

Theorem 3. *There is an infinite sequence of distinct integers n such that there are languages L'_n such that there are quantum finite automata with mixed states with $5n$ states recognizing the language L'_n with probability $\frac{3}{4}$ while any deterministic finite automaton recognizing L'_n needs to have at least $e^{O(n \ln n)}$ states.*

Proof is delayed till Section 4.

Since the number of the states for deterministic automata and quantum automata with pure states differ no more than exponentially, we have

Theorem 4. *There is an infinite sequence of distinct integers n such that there are languages L_n in a 2-letter alphabet such that there are quantum finite automata with mixed states with $2n$ states recognizing the language L_n with probability $\frac{3}{4}$ while any quantum finite automaton with pure states recognizing L_n with bounded error needs to have at least $e^{O(n \ln n)}$ states.*

Unfortunately, the alphabet for the languages considered in Theorem 3 grows unlimitedly with n. It is only natural to try to prove a counterpart of Theorem 3 in 2- or 3-letter alphabet. We have developed a methodology of such a proof based on combining ideas of Theorem 3 and the results in [10]. However we need a notion similar to Hamming codes for permutations. Since Hamming distance between permutations is already considered in several well-known textbooks (e.g. [7]) it seemed natural that the corresponding theory might be already published. Very far from truth!

2   Permutations

Permutation of the set N_n is a 1-1 correspondence from N_n onto itself. Let f be such a permutation. The fact that it is *onto* means that for any $k \in N_n$ there exists $i \in N_n$ such that $f(i) = k$.

If we think of a permutation that "changes" the position of the first element to the first element, the second to the second, and so on, we really have not changed the positions of the elements at all. Because of its action, we describe it as the identity permutation because it acts as an identity function.

There are two main notations for such permutations. In relation notation, one can just arrange the "natural" ordering of the elements being permuted on a row, and the new ordering on another row:

$$\left\{ \begin{matrix} 1\ 2\ 3\ 4\ 5 \\ 2\ 5\ 4\ 3\ 1 \end{matrix} \right\}$$

stands for the permutation s of the set $\{1, 2, 3, 4, 5\}$ defined by $s(1) = 2, s(2) = 5, s(3) = 4, s(4) = 3, s(5) = 1$.

Rather often in the literature permutations are described by the string $s(1) = 2, s(2) = 5, s(3) = 4, s(4) = 3, s(5) = 1$ only. Alternatively, we can write the permutation in terms of how the elements change when the permutation is successively applied. This is referred to as the permutation's decomposition in a product of disjoint cycles. It works as follows: starting from one element x, we write the sequence $(x s(x) s^2(x) \cdots)$ until we get back the starting element (at which point we close the parenthesis without writing it for a second time). This is called the cycle associated to x's orbit following s. Then we take an element we did not write yet and do the same thing, until we have considered all elements. In the above example, we get: $s = (125)(34)$.

Every fixed point is a cycle with length 1.

If we have a finite set E of n elements, it is by definition in bijection with the set $1, ..., n$, where this bijection f corresponds just to numbering the elements. Once they are numbered, we can identify the permutations of the set E with permutations of the set $\{1, ..., n\}$.

If one has some permutation, called P, one may describe a permutation, written P^{-1}, which undoes the action of applying P. In essence, performing P then P^{-1} is equivalent to performing the identity permutation. One always has such a permutation since a permutation is a bijective map. Such a permutation is called the inverse permutation.

One can define the product of two permutations. If we have two permutations, P and Q, the action of performing P and Q will be the same as performing some other permutation, $R = P \circ Q$, itself. The product of P and Q is defined to be the permutation R. An even permutation is a permutation which can be expressed as the product of an even number of transpositions, and the identity permutation is an even permutation as it equals $(12) \circ (12)$. An odd permutation is a permutation which can be expressed as the product of an odd number of transpositions. It can be shown that every permutation is either odd or even and cannot be both.

The set of all permutations of the set $1, ..., n$ with algebraic operation "product of permutations" can be considered as a group G_n. This group has two generating elements, the permutations $(123 \cdots n)$ and $(12)(3)(4) \cdots (n)$.

We can also represent a permutation in matrix form - the resulting matrix is known as a permutation matrix.

A permutation matrix is a matrix obtained by permuting the rows of an $n \times n$ identity matrix according to some permutation of the numbers 1 to n. Every row and column therefore contains precisely a single 1 with 0s everywhere else, and every permutation corresponds to a unique permutation matrix. There are therefore $n!$ permutation matrices of size n, where $n!$ is a factorial.

3   Hamming Distance

Hamming distance between two objects is the number of changes one needs to perform to obtain one object from the another. The Hamming distance between two binary words (of the same length) is defined to be the number of positions at which they differ. For instance, we consider a set of three binary words $\{0011, 0110, 1100\}$. The first word is at Hamming distance 3 from the other two. Additionally, every word in this set is at Hamming distance at least 2 from any other. Such systems of words are called codes. They are important because they allow us to eliminate accidental errors when transmitting the words through noisy information channels.

We consider Hamming distance between permutations. Hamming distance between the permutation s of the set $\{1, 2, 3, \cdots n\}$ and the permutation r of the same set is the number of distinct numbers i such that $s(i) \neq r(i)$. For instance, let s be a permutation of the set $\{1, 2, 3, \cdots n\}$ and the number of it's fixed points be p. Then the Hamming distance between the permutation s and the identity permutation is the number $n - p$.

It would be interesting to develop a theory of Hamming codes for permutations, i.e. sets of permutations such that any two distinct permutations in the set have Hamming distance at least d. Unfortunately, we were not able to find in the literature a solution to this problem. The difficulty is in the fact that in the traditional Hamming codes for binary strings positions in the string are independent while positions in a permutation are not independent. For instance, any two permutations of the same set either coinside or their Hamming distance is at least 2.

For arbitrary n there is a set P_3 of n-permutations such that the Hamming distance between any two distinct permutations in this set is at least 3. Take the set of all even permutations as P_3. It is easy to see that there is no bigger set of n-permutations with this property.

For arbitrary n there is a set P_n of n-permutations such that the Hamming distance between any two distinct permutations in this set is at least n. Take the set of all cyclic permutations of type $s = x + d(mod\, n)$ where $d \in \{0, 1, 2, \cdots, n-1\}$ as P_n.

It is easy to see that there is no bigger set of n-permutations with this property. It is easy to see that both P_3 and P_n are groups with the operation "product of permutations". It is more difficult to construct maximum cardinality sets P_d for d between 3 and n. We have only partial results for this. However, our main goal is the complexity of quantum finite automata, not permutations. The subsequent sections contain results on Hamming distance between permutations sufficient for our goal.

Lemma 1. *Let d be an arbitrary real number such that $0 \leq d \leq 1$. No more than $2^{dn\ln n}$ permutations can be on Hamming distance less or equal than $(dn$ from the identity permutation.*

Proof. By Stirling formula, $n! = e^{n.\ln n - o(n\ln n)}$. Let π be an arbitrary n-permutation. How many there are distinct n-permutations differing from the permutation π in no more than dn positions? The differing positions can be chosen in

$$\leq \binom{n}{d} < 2^n$$

ways and these $\leq dn$ positions are permuted. Hence there are no more than $2^n . 2^{dn\ln n - o(n\ln n)} \leq 2^{dn\ln n}$ permutations of this type.

Theorem 5. *For arbitrary constant $c < 1$ such that for arbitrary n there is a set G_n of n-permutations containing $e^{\Omega(n \log n)}$ permutations such that the pairwise Hamming distance of permutations is at least $c \cdot n$.*

Proof. Immediately from Lemma 1.

4   Main Results on Automata

Theorem 6. *The assertion (1) implies the assertion (2), where:*

(1) there is a fixed constant c and an infinite sequence of distinct integers n such that for each n there is a group G_n of permutations of the set $\{1, 2, \ldots, n\}$, the group has $e^{\Omega(n \log n)}$ elements and k generating elements, and the pairwise Hamming distance of permutations is at least $c \cdot n$,

(2) there is an infinite sequence of distinct integers n such that for each n there is a language L_n in a k-letter alphabet that can be recognized with probability $\frac{c}{2}$ by a quantum finite automata with mixed states that has $2n$ states, while any deterministic finite automaton recognizing L_n must have at least $e^{\Omega(n \log n)}$ states.

Proof. For each permutation group G_n we define the language L_n as follows:

> The letters of L_n are the k generators of the group G_n and
> it consists of words $s_1 s_2 s_3 \ldots s_m$ such that the product
> $s_1 \circ s_2 \circ s_3 \circ \cdots \circ s_m$ differs from the identity permutation.

Proof. (1) Any deterministic automaton recognizing L_n is to remember the first input letter by a specific state.

(2) We will construct a quantum automaton with mixed states. It has $4n$ states and the initial density matrix ρ_0 is a diagonal block-matrix that consists of n blocks $\tilde{\rho}_0$:

$$\tilde{\rho}_0 = \frac{1}{2n} \begin{pmatrix} 1 & 1 & 0 & 0 \\ 1 & 1 & 0 & 0 \\ 0 & 0 & 0 & 0 \\ 0 & 0 & 0 & 0 \end{pmatrix}$$

For each of k generators $g_i \in G_n$ we will construct the corresponding unitary matrix U_i as follows – it is a $2n \times 2n$ permutation matrix, that permutes the elements in the even positions according to permutation g_i, but leaves the odd positions unpermuted.

For example, $g = 3241$ can be expressed as the following permutation matrix that acts on a column vector:

$$g = \begin{pmatrix} 0 & 0 & 1 & 0 \\ 0 & 1 & 0 & 0 \\ 0 & 0 & 0 & 1 \\ 1 & 0 & 0 & 0 \end{pmatrix}$$

The initial density matrix ρ_0 for $n = 4$ and the unitary matrix U that corresponds to the permutation matrix (4) of permutation g are as follows:

$$\rho_0 = \frac{1}{8} \begin{pmatrix} 1\,1\,0\,0\,0\,0\,0\,0 \\ 1\,1\,0\,0\,0\,0\,0\,0 \\ 0\,0\,1\,1\,0\,0\,0\,0 \\ 0\,0\,1\,1\,0\,0\,0\,0 \\ 0\,0\,0\,0\,1\,1\,0\,0 \\ 0\,0\,0\,0\,1\,1\,0\,0 \\ 0\,0\,0\,0\,0\,0\,1\,1 \\ 0\,0\,0\,0\,0\,0\,1\,1 \end{pmatrix}$$

$$U = \begin{pmatrix} 1\,0\,0\,0\,0\,0\,0\,0 \\ 0\,0\,0\,0\,0\,1\,0\,0 \\ 0\,0\,1\,0\,0\,0\,0\,0 \\ 0\,0\,0\,1\,0\,0\,0\,0 \\ 0\,0\,0\,0\,1\,0\,0\,0 \\ 0\,0\,0\,0\,0\,0\,0\,1 \\ 0\,0\,0\,0\,0\,0\,1\,0 \\ 0\,1\,0\,0\,0\,0\,0\,0 \end{pmatrix}$$

The unitary matrix $U_\$$ for the end-marker is also a diagonal block-matrix. It consists of n blocks that are the *Hadamard matrices*

$$\tilde{H} = \frac{1}{\sqrt{2}} \begin{pmatrix} 1 & 1 \\ 1 & -1 \end{pmatrix}$$

Notice how the Hadamard matrix $\tilde{H}$ acts on two specific 2×2 density matrices:

$$\text{if } \rho = \frac{1}{2n} \begin{pmatrix} 1 & 1 \\ 1 & 1 \end{pmatrix}, \text{ then } \tilde{H}\rho\tilde{H}^\dagger = \frac{1}{2n} \begin{pmatrix} 2 & 0 \\ 0 & 0 \end{pmatrix},$$

$$\text{if } \rho = \frac{1}{2n} \begin{pmatrix} 1 & 0 \\ 0 & 1 \end{pmatrix}, \text{ then } \tilde{H}\rho\tilde{H}^\dagger = \frac{1}{2n} \begin{pmatrix} 1 & 0 \\ 0 & 1 \end{pmatrix}.$$

For example, when the letter g is read, the unitary matrix U is applied to the density matrix ρ_0 (both are given in equation (4)) and the density matrix $\rho_1 = U\rho_0 U^\dagger$ is obtained. When the end-marker "\$" is read, the density matrix becomes $\rho_\$ = U_\$ \rho_1 U_\$^\dagger$. Matrices ρ_1 and $\rho_\$$ are as follows:

$$\rho_1 = \frac{1}{8} \begin{pmatrix} 1\,0\,0\,0\,0\,0\,0\,1 \\ 0\,1\,0\,0\,1\,0\,0\,0 \\ 0\,0\,1\,1\,0\,0\,0\,0 \\ 0\,0\,1\,1\,0\,0\,0\,0 \\ 0\,1\,0\,0\,1\,0\,0\,0 \\ 0\,0\,0\,0\,0\,1\,1\,0 \\ 0\,0\,0\,0\,0\,1\,1\,0 \\ 1\,0\,0\,0\,0\,0\,0\,1 \end{pmatrix}, \rho_\$ = \frac{1}{8} \begin{pmatrix} 1 & 0 & 0 & 0 & \frac{1}{2} & \frac{1}{2} & \frac{1}{2} & -\frac{1}{2} \\ 0 & 1 & 0 & 0 & -\frac{1}{2} & -\frac{1}{2} & \frac{1}{2} & -\frac{1}{2} \\ 0 & 0 & 2 & 0 & 0 & 0 & 0 \\ 0 & 0 & 0 & 0 & 0 & 0 & 0 \\ \frac{1}{2} & -\frac{1}{2} & 0 & 0 & 1 & 0 & \frac{1}{2} & \frac{1}{2} \\ \frac{1}{2} & -\frac{1}{2} & 0 & 0 & 0 & 1 & -\frac{1}{2} & -\frac{1}{2} \\ \frac{1}{2} & \frac{1}{2} & 0 & 0 & \frac{1}{2} & -\frac{1}{2} & 1 & 0 \\ -\frac{1}{2} & -\frac{1}{2} & 0 & 0 & \frac{1}{2} & -\frac{1}{2} & 0 & 1 \end{pmatrix}.$$

Finally, we declare the states in the even positions to be accepting, but the states in the odd positions to be rejecting. Therefore one must sum up the diagonal entries that are in the even positions of the final density matrix to find the probability that a given word is accepted.

In our example the final density matrix $\rho_\$$ is given in (4). It corresponds to the input word "$g\$$", which is accepted with probability $\frac{1}{8}(1 + 0 + 1 + 1) = \frac{3}{8}$ and rejected with probability $\frac{1}{8}(1 + 2 + 1 + 1) = \frac{5}{8}$. Note that the accepting and rejecting probabilities sum up to 1.

It is easy to see, that the words that do not belong to the language L_n are rejected with certainty, because the matrix $U_\$ \rho_0 U_\$^\dagger$ has all zeros in the even positions on the main diagonal. However, the words that belong to L_n are accepted with the probability at least $\frac{d}{2n} = \frac{cn}{2n} = \frac{c}{2}$, because all permutations are at least at the distance d from the identity permutation.

It is also easy to see that any deterministic automaton that recognizes the language L_n must have at least $N = |G_n|$ states, where $|G_n|$ is the size of the permutation group G_n. If the number of states is less than N, then there are two distinct words u and v such that the deterministic automaton ends up in the same state no matter which one of the two words it reads. Since G_n is a group, for each word we can find an inverse, that returns the automaton in the initial state (the only rejecting state). Since u and v are different, they have different inverses and $u \circ u^{-1}$ is the identity permutation and must be rejected, but $v \circ u^{-1}$ is not the identity permutation and must be accepted – a contradiction.   □

5   Super-Exponential Size Advantage

Finally, we wish to prove Theorem 3.

Consider the following infinite sequence of languages. For every n take the set G_n considered in Theorem 5. The language L'_n consists of all the words aa (of the length 2) where a is a symbol for an arbitrary element from G_n. Hence there are $e^{\Omega(n \log n)}$ letters in the alphabet of the language L'_n and equally many words in L'_n.

Theorem 3. There is an infinite sequence of distinct integers n such that there are languages L'_n such that there are quantum finite automata with mixed states with $5n$ states recognizing the language L'_n with probability $\frac{3}{4}$ while any deterministic finite automaton recognizing L'_n needs to have at least $e^{O(n \ln n)}$ states.

Proof is similar to the proof of Theorem 6. When constructing the quantum automaton for Theorem 6 we used two distinct sets of states $\{q_1, q_3, \cdots, q_{2n-1}\}$ and $\{q_2, q_4, \cdots, q_{2n}\}$. The unitary transformations corresponding to all generators in the group G_n permuted some states in the first set and it left all the states in the second set unpermuted.

Now we have five such sets of n-tuples of states. Let g_i be any one of the permutations in G_n. (Remember that the cardinality of G_n equals $e^{\Omega(n \log n)}$.)

Let the inverse permutation of g_i be denoted as h_i. The unitary transformation corresponding to g_i leaves the states from the first set unpermuted. It takes the states from the second set into the third one performing the permutation g_i. For instance, if $n = 5$ and g_i is (35421) then $f(6) = 13, f(7) = 15, f(8) = 14, f(9) = 12, f(10) = 11$. The unitary transformation takes the third set into the fourth one performing the permutation h_i. The unitary transformation takes the fourth set into the fifth one performing no permutation. The unitary transformation takes the fifth set into the first one performing no permutation. After this transformation a measurement is performed which measures all the states in the fifth set considering them as rejecting states. No measurement of accepting states is performed. The unitary transformation corresponding to the end-marker uses n instances of the *Hadamard matrices*

$$\tilde{H} = \frac{1}{\sqrt{2}} \begin{pmatrix} 1 & 1 \\ 1 & -1 \end{pmatrix}.$$

This transformation has the following property. If the state q_{3n+j} has been the result of permutation from the state q_{n+j}, then the amplitude of the state q_{4n+j} becomes double the amplitude of the state q_{3n+j} on the preceding step. If the state q_{3n+j} has been the result of permutation from the state q_{n+j}, then the amplitude of the state q_{4n+j} becomes equal to the amplitude of the state q_{3n+j} on the preceding step. After this transformation a measurement is performed which measures all the states in the first set considering them as accepting states. All the other states are measured as rejecting states.

It is easy to see that if the input word is aa, then all the amplitudes of the states q_{3n+j} are doubled at the moment when the end-marker is read. The word is accepted with the probability 1. If the input word is shorter than two letters, then the probability to accept this word equals zero. If the input word is longer than two letters, then the probability of acceptance cannot exceed $\frac{1}{2}$. If the length of the input word is two letters but the letters are not equal, then by the property of G_n desribed in Theorem 5, the probability of acceptance is less than $\frac{1}{2}$.

6   Smaller Alphabets

Unfortunately, the size of the alphabet for the languages considered in Theorem 3 grow extraordinary quickly. What can we prove in the case of 2- or 3-letter alphabets? The paper [11] by R.Freivalds, M.Ozols and L.Mančinska shows a large collection of examples found in order to prove

Conjecture. There is a fixed constant c and an infinite sequence of distinct integers n such that for each n there is a group G_n of permutations of the set $\{1, 2, \ldots, n\}$, the group has $e^{\Omega(n \log n)}$ elements and k generating elements, and the pairwise Hamming distance of permutations is at least $c \cdot n$.

The parameters of some of these examples may be seen at this Figure.

Some of these examples are constructed in [11] by computerised search.

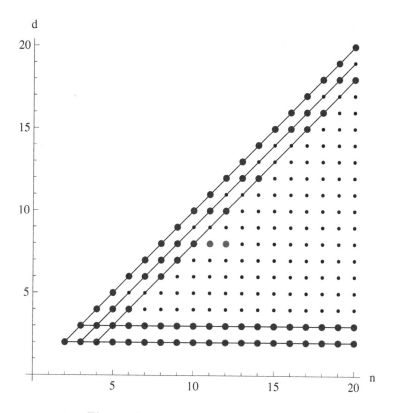

Fig. 1. The maximal permutation groups

We performed computer experiments to find permutation groups with pairwise Hamming distance in the region between $d \geq 4$ and $d \leq n - 3$. The obtained results for $n = 7, 8, 9, 10$ are shown in Table 1. In addition we mention also two large groups for $n = 15$ and $n = 16$.

These groups were obtained by choosing two random permutations g_1 and g_2 and computing their closure with respect to the product of permutations. If at some point the distance between any two distinct obtained permutations became less than some predefined d_{min}, the process was terminated and restarted with another random generators g_1 and g_2. Some of the groups obtained in this way have very interesting properties:

(1) $G(7, 4)$ has $168 = 7 \cdot 6 \cdot 4$ elements and is isomorphic to the automorphism group of the *Fano plane*.

(2) $G(8, 4)$ has $1344 = 8 \cdot 168 = 8 \cdot 7 \cdot 6 \cdot 4$ elements. This group has the property, that the stabilizers of any element form a group that is isomorphic to the automorphism group of the Fano plane. This group also has a property that for any 3-tuples x and y of distinct elements there are exactly 4 permutations that send x to y. It is isomorphic to the automorphism group of the *octonion* multiplication table.

Table 1. Experimentally obtained results for $G(n, d)$. The columns have the following meaning: n – the size of the set S, d – the pairwise Hamming distance, $G(n, d)$ – the size of the group obtained, "Bound" – the upper bound for $G(n, d)$, "Generators" – the two generators of the group.

n	d	$G(n, d)$	Bound	Generators
7	4	168	840	$6, 4, 3, 2, 5, 1, 7$
				$6, 1, 7, 5, 2, 3, 4$
8	5	336	1680	$3, 8, 6, 2, 4, 5, 1, 7$
				$7, 4, 6, 3, 1, 5, 2, 8$
8	4	1344	6720	$2, 6, 8, 4, 5, 7, 1, 3$
				$7, 4, 3, 5, 1, 8, 6, 2$
9	6	1512	3024	$4, 5, 1, 8, 3, 7, 6, 2, 9$
				$3, 4, 8, 5, 7, 1, 6, 9, 2$
9	5	1512	15120	$9, 4, 1, 6, 5, 2, 7, 8, 3$
				$1, 4, 5, 3, 7, 9, 8, 2, 6$
9	4	1512	60480	$7, 2, 8, 3, 5, 6, 9, 4, 1$
				$6, 1, 3, 8, 2, 4, 9, 5, 7$
10	7	720	5040	$3, 9, 5, 7, 4, 8, 10, 6, 1, 2$
				$7, 9, 4, 5, 3, 6, 8, 1, 10, 2$
10	6	1512	30240	$8, 2, 10, 7, 4, 3, 1, 6, 5, 9$
				$1, 2, 8, 5, 10, 6, 3, 7, 9, 4$
10	5	1512	151200	$1, 10, 3, 9, 6, 8, 5, 4, 7, 2$
				$1, 10, 8, 3, 2, 4, 5, 7, 6, 9$
10	4	1920	604800	$5, 1, 4, 8, 9, 7, 6, 10, 2, 3$
				$7, 8, 2, 1, 10, 3, 9, 6, 4, 5$
15	12	2520	32760	$7, 2, 4, 5, 11, 10, 13, 15, 3, 9, 6, 8, 14, 12, 1$
				$9, 15, 11, 6, 4, 2, 10, 13, 7, 12, 8, 1, 14, 3, 5$
16	12	40320	524160	$16, 5, 6, 12, 14, 13, 11, 1, 10, 3, 7, 4, 15, 8, 9, 2$
				$6, 7, 14, 8, 15, 3, 12, 2, 9, 10, 13, 11, 4, 16, 1, 5$

(3) $G(8, 4)$ has $1512 = 9 \cdot 168 = 9 \cdot 8 \cdot 7 \cdot 3$ elements and it has the same stabilizer property, but for each 3-tuples x and y there are exactly 3 permutations that send x to y.

(4) $G(15, 12)$ has $2520 = 15 \cdot 168 = 15 \cdot 14 \cdot 12$ elements and it also has the stabilizer property, but for each 2-tuples x and y there are exactly 12 permutations that send x to y.

(5) $G(16, 12)$ has $40320 = 16 \cdot 15 \cdot 168 = 16 \cdot 15 \cdot 14 \cdot 12$ elements. The stabilizers of any two elements form a group that is isomorphic to the automorphism group of the Fano plane. For any 3-tuples x and y there are exactly 12 permutations that send x to y.

Acknowledgments

We would like to acknowledge Simone Severini for drawing our attention to the Mathieu groups and for his efforts trying to prove the Conjecture above in Section 6.

References

1. Aharonov, D., Kitaev, A., Nisan, N.: Quantum circuits with mixed states. In: Proc. STOC 1998, pp. 20-30 (1998)
2. Ambainis, A.: The complexity of probabilistic versus deterministic finite automata. LNCS, vol. 1178, pp. 233–237. Springer, Heidelberg (1996)
3. Ambainis, A., Beaudry, M., Golovkins, M., Kikusts, A., Mercer, M., Thérien, D.: Algebraic Results on Quantum Automata. Theory Comput. Syst. 39(1), 165–188 (2006)
4. Ambainis, A., Freivalds, R.: 1-way quantum finite automata: strengths, weaknesses and generalizations. Proc. IEEE FOCS'98 , 332–341 (1998)
5. Ambainis, A., Barbans, U., Belousova, A., Belovs, A., Dzelme, I., Folkmanis, G., Freivalds, R., Ledins, P., Opmanis, R., Škuškovniks, A.: Size of Quantum Versus Deterministic Finite Automata. VLSI 2003 , 303–308 (2003)
6. Artin, E.: Beweis des allgemeinen Reziprozitätsgesetzes. Mat. Sem. Univ. Hamburg 5, 353–363 (1927)
7. Cameron, P.: Permutation groups. London Mathematical Society Student Texts series. Cambridge University Press, Cambridge (1999)
8. Freivalds, R.: On the growth of the number of states in result of the determinization of probabilistic finite automata. Avtomatika i Vichislitel'naya Tekhnika (Russian) 3, 39–42 (1982)
9. Freivalds, R.: Languages Recognizable by Quantum Finite Automata. In: Farré, J., Litovsky, I., Schmitz, S. (eds.) CIAA 2005. LNCS, vol. 3845, pp. 1–14. Springer, Heidelberg (2006)
10. Freivalds, R.: Non-constructive methods for finite probabilistic automata. In: Proceedings of the 11th International Conference Developments in Language Theory DLT-2007, Turku, Finland, July 3-5, 2007. LNCS, vol. 4588, Springer, Heidelberg (to be published, 2007)
11. Freivalds, R., Ozols, M., Mančinska, L.: Permutation Groups and the Strength of Quantum Finite Automata with Mixed States. In: Proceedings of the Workshop "Probabilistic and Quantum Algorithms" colocated with the 11th International Conference Developments in Language Theory DLT-2007, Turku, Finland, July 3-5, 2007 (to be published, 2007)
12. Kondacs, A., Watrous, J.: On the power of quantum finite state automata. In: Proc. IEEE FOCS'97, pp. 66–75 (1997)

Probabilistic Techniques in Algorithmic Game Theory[*]
(SAGA 2007 Invited Paper)

Spyros C. Kontogiannis[1,2] and Paul G. Spirakis[2]

[1] Computer Science Department, University of Ioannina,
45110 Ioannina, Greece
kontog@cs.uoi.gr
[2] Research Academic Computer Technology Institute,
P.O. Box 1382, N. Kazantzaki Str., 26500 Rio–Patra, Greece
{kontog,spirakis}@cti.gr

Abstract. We consider applications of probabilistic techniques in the framework of algorithmic game theory. We focus on three distinct case studies: (i) The exploitation of the probabilistic method to demonstrate the existence of approximate Nash equilibria of logarithmic support sizes in bimatrix games; (ii) the analysis of the statistical conflict that mixed strategies cause in network congestion games; (iii) the effect of coalitions in the quality of congestion games on parallel links.

Keywords: Game Theory, Atomic Congestion Games, Coalitions, Convergence to Equilibria, Price of Anarchy.

1 Preliminaries and Notation

For any $k \in \mathbb{N}$, let $[k] \equiv \{1, 2, \ldots, k\}$. $M \in F^{m \times n}$ denotes a $m \times n$ matrix (denoted by capital letters) whose elements belong to set F. We call a pair $(A, B) \in (F \times F)^{m \times n}$ (ie, an $m \times n$ matrix whose elements are *ordered pairs* of values from F) a **bimatrix**. A $k \times 1$ matrix is also considered to be an k-**vector**. Vectors are denoted by bold small letters (eg, $\mathbf{x}$). $\mathbf{e_i}$ denotes a vector having a 1 in the i-th position and 0 everywhere else. $\mathbf{1_k}$ ($\mathbf{0_k}$) is the k-vector having 1s (0s) in all its coordinates. The $k \times k$ matrix $E = \mathbf{1_k} \cdot \mathbf{1_k}^T \in \{1\}^{k \times k}$ has value 1 in all its elements. For any $\mathbf{x}, \mathbf{y} \in \mathbb{R}^n$, $\mathbf{x} \geq \mathbf{y}$ implies their component–wise comparison: $\forall i \in [n], x_i \geq y_i$. For any $m \times n$ (bi)matrix M, M_j is its j-th column (as an $m \times 1$ vector), M^i is the i-th row (as a (transposed) $1 \times n$ vector) and $M_{i,j}$ is the (i, j)-th element.

For any integer $k \geq 1$, $\Delta_k = \{\mathbf{z} \in \mathbb{R}^k : \mathbf{z} \geq \mathbf{0}; \ (\mathbf{1_k})^T \mathbf{z} = 1\}$ is the $(k - 1)$-simplex. For any point $\mathbf{z} \in \Delta_k$, its **support** is the set of coordinates with positive value: $supp(\mathbf{z}) \equiv \{i \in [k] : z_i > 0\}$. For an arbitrary logical expression $\mathcal{E}$, we denote by $\mathbb{P}\{\mathcal{E}\}$ the probability of this expression being true, while $\mathbb{I}_{\{\mathcal{E}\}}$ is the indicator variable of whether $\mathcal{E}$ is true or false. For any random variable x, $\mathbb{E}\{x\}$ is its expected value (wrt some probability measure).

---

[*] Partially supported by EU/6th Framework Programme, contract 001907 (DELIS).

J. Hromkovič et al. (Eds.): SAGA 2007, LNCS 4665, pp. 30–53, 2007.
© Springer-Verlag Berlin Heidelberg 2007

2   Existence of Approximate Equilibria of Small Support Sizes in Bimatrix Games

In this section we use the probabilistic method in order to prove the existence of (well supported) approximate Nash equilibria of logarithmic support sizes in bimatrix games. In the next subsection we briefly present the model of bimatrix games and the notions of approximate equilibria for these games. Consequently we demonstrate the proof of Althöfer's Approximation Lemma and how this lemma can be used to prove existence of well supported approximate equilibria with small support sizes.

2.1   Bimatrix Games Notation

For $2 \le m \le n$, an $m \times n$ **bimatrix game** $\langle A, B \rangle$ is a $2-$person game in normal form, determined by the bimatrix $(A, B) \in (\mathbb{R} \times \mathbb{R})^{m \times n}$ as follows: The first player (the **row player**) has an $m-$element *action set* $[m]$, and the second player (the **column player**) has an $n-$element *action set* $[n]$. Each row (column) of the bimatrix corresponds to a different action of the row (column) player. The row and the column player's payoffs are determined by the $m \times n$ real matrices A and B respectively. $\langle A, B \rangle$ is a **zero sum** game, if $B = -A$. In that case the game is solvable in polynomial time using linear programming. If both payoff matrices belong to $[0, 1]^{m \times n}$ then we have a $[0, 1]-$**bimatrix** (aka **normalized**) **game**. The special case of bimatrix games in which all elements of the bimatrix belong to $\{0, 1\} \times \{0, 1\}$, is called a $\{0, 1\}-$**bimatrix** (aka **win lose**) **game**. A win lose game having (for integer $\lambda \ge 1$) at most λ $(1, 0)-$elements per row and at most λ number $(0, 1)-$element per column of the bimatrix, is called a $\lambda-$**sparse** game. Any point $\mathbf{x} \in \Delta_m$ is a **mixed strategy** for the row player: She determines her action independently from the column player, according to the probability vector $\mathbf{x}$. Similarly, any point $\mathbf{y} \in \Delta_n$ is a mixed strategy for the column player. Each extreme point $\mathbf{e_i} \in \Delta_m$ ($\mathbf{e_j} \in \Delta_n$), that enforces the use of the i-th row (j-th column) by the row (column) player, is a **pure strategy** for her. Any element $(\mathbf{x}, \mathbf{y}) \in \Delta_m \times \Delta_n$ is a (mixed in general) **strategy profile** for the players. The notion of *approximate best responses* will help us simplify the forthcoming definitions:

Definition 1 (Approximate   Best   Response).   *Fix   any   bimatrix   game $\langle A, B \rangle$ and $0 \le \varepsilon \le 1$.   The   sets   of   **approximate (pure) best responses**   of   the   column   (row)   player   against $\mathbf{x} \in \Delta_m$ ($\mathbf{y} \in \Delta_n$)   are: $BR(\varepsilon, B^T, \mathbf{x}) = \{\mathbf{y} \in \Delta_n : \mathbf{y}^T B^T \mathbf{x} \ge \mathbf{z}^T B^T \mathbf{x} - \varepsilon, \forall \mathbf{z} \in \Delta_n\}$, $BR(\varepsilon, A, \mathbf{y}) = \{\mathbf{x} \in \Delta_m : \mathbf{x}^T A \mathbf{y} \ge \mathbf{z}^T A \mathbf{y} - \varepsilon, \forall \mathbf{z} \in \Delta_m\}$, $PBR(\varepsilon, B^T, \mathbf{x}) = \{j \in [n] : B_j^T \mathbf{x} \ge B_s^T \mathbf{x} - \varepsilon, \forall s \in [n]\}$ and $PBR(\varepsilon, A, \mathbf{y}) = \{i \in [m] : A^i \mathbf{y} \ge A^r \mathbf{y} - \varepsilon, \forall r \in [m]\}$.*

Definition 2 (Approximate Nash Equilibria). *For any bimatrix game $\langle A, B \rangle$ and $0 \le \varepsilon \le 1$, $(\mathbf{x}, \mathbf{y}) \in \Delta_m \times \Delta_n$ is: (1) An $\varepsilon-$**approximate Nash***

Equilibrium *(ε−ApproxNE) iff each player chooses an ε−approximate best response against the opponent:* $[\mathbf{x} \in BR(\varepsilon, A, \mathbf{y})] \wedge [\mathbf{y} \in BR(\varepsilon, B^T, \mathbf{x})]$. *(2) An* ε−**well–supported Nash Equilibrium** *(ε−SuppNE) iff each player assigns positive probability only to ε−approximate* pure best responses *against the strategy of the opponent:* $\forall i \in [m], x_i > 0 \Rightarrow i \in PBR(\varepsilon, A, \mathbf{y})$ *and* $\forall j \in [n], y_j > 0 \Rightarrow j \in PBR(\varepsilon, B^T, \mathbf{x})$.

To see the difference between the two notions of approximate equilibria, consider the Matching Pennies game with payoffs $(A, B)_{1,1} = (A, B)_{2,2} = (1, 0)$ (row player wins) and $(A, B)_{1,2} = (A, B)_{2,1} = (0, 1)$ (column player wins). $\left(\mathbf{e_1}, \frac{1}{2} \cdot (\mathbf{e_1} + \mathbf{e_2})\right)$ is a 0.5−ApproxNE, but only a 1−SuppNE for the game.

Any NE is both a 0−ApproxNE and a 0−SuppNE. Moreover, any ε−SuppNE is also an ε−ApproxNE, but *not necessarily vice versa*, as was shown in the previous example. Indeed, the only thing we currently know towards this direction is that from an arbitrary $\frac{\varepsilon^2}{8n}$−ApproxNE one can construct an ε−SuppNE in polynomial time [6]. Note that both notions of approximate equilibria are defined wrt an *additive* error term ε. Although (exact) NE are known not to be affected by any positive scaling of the payoffs, approximate notions of NE are indeed affected. Therefore, from now on we adopt the commonly used assumption in the literature (eg, [21,8,17,5,6]) that, when referring to ε−ApproxNE or ε−SuppNE, we consider a [0, 1]−bimatrix game.

Definition 3 (Uniform Profiles). $\mathbf{x} \in \Delta_r$ *is a* k−**uniform strategy** *iff* $\mathbf{x} \in \Delta_r(k) = \Delta_r \cap \left\{0, \frac{1}{k}, \frac{2}{k}, \ldots, \frac{k-1}{k}, 1\right\}^r$. *If* $\mathbf{x} \in \hat{\Delta}_r(k) = \Delta_r \cap \left\{0, \frac{1}{k}\right\}^r$, *then we refer to a* **strict** k−**uniform strategy**. $(\mathbf{x}, \mathbf{y}) \in \Delta_m \times \Delta_n$ *for which* $\mathbf{x}$ *is a (strict)* k−*uniform strategy and* $\mathbf{y}$ *is a (strict)* ℓ−*uniform strategy, is called a* **(strict)** (k, ℓ)−**uniform** *profile.*

2.2   Existence of SuppNE with Logarithmic Support Sizes

The existence of (uniform) ε−ApproxNE with small support sizes was proved in [21]. Recently [18] extended this to ε−SuppNE, based solely on Althöfer's *Approximation Lemma* [1], which is a direct application of the probabilistic method:

Theorem 1 (Approximation Lemma [1]). *Consider any real matrix* $C \in [0, 1]^{m \times n}$. *Let* $\mathbf{p} \in \Delta_m$ *be any probability vector. Fix an arbitrary constant* $\varepsilon > 0$. *There exists probability vector* $\hat{\mathbf{p}} \in \Delta_m$ *with* $|supp(\hat{\mathbf{p}})| \leq k \equiv \left\lceil \frac{log(2n)}{2\varepsilon^2} \right\rceil$, *such that* $|\mathbf{p}^T C_j - \hat{\mathbf{p}}^T C_j| \leq \varepsilon$, $\forall j \in [n]$. *Moreover,* $\hat{\mathbf{p}}$ *is a* k−*uniform strategy, ie,* $\hat{\mathbf{p}} \in \Delta_m(k)$.

Proof. Fix $C \in [0, 1]^{m \times n}$ and consider arbitrary probability vector $\mathbf{p} \in \Delta_m$. For some integer k (to be specified later), consider the independent random variables $X_1, X_2, \ldots, X_k$ taking values from $[m]$, according to $\mathbf{p}$, ie, $\forall t \in [k], \forall i \in [m], \mathbb{P}\{X_t = i\} = p_i$.

Fix now an arbitrary column $j \in [n]$ of C and let $\forall t \in [k], Y_t(j) = C_{X_t, j}$ be *independent* random variables taking values in $\{C_{i,j}\}_{i \in [m]} \subset [0, 1]$. Let $\bar{Y}(j) =$

$\frac{1}{k}\sum_{t\in[k]}Y_t(j) = \frac{1}{k}\sum_{t\in[k]}C_{X_t,j}$ be the average value of $\{Y_t(j)\}_{t\in[k]}$. Observe that $\forall t \in [k], \mathbb{E}\{Y_t(j)\} = \sum_{i=1}^{m}p_i\cdot C_{i,j} = \mathbf{p}^T C_j$ and, by linearity of expectation, $\mathbb{E}\{\bar{Y}(j)\} = \mathbf{p}^T C_j$ as well. Using the Hoeffding Bound [15] and the Union Bound we conclude that: $\forall\varepsilon > 0$,

$$\forall j \in [n], \ \mathbb{P}\{|\bar{Y}(j) - \mathbf{p}^T C_j| > \varepsilon\} \leqslant 2\exp(-2\varepsilon^2 k) \quad \overset{/* \text{ Union Bound } */}{\Longrightarrow}$$

$$\mathbb{P}\{\exists j \in [n] : |\bar{Y}(j) - \mathbf{p}^T C_j| > \varepsilon\} \leqslant 2n\exp(-2\varepsilon^2 k) = \exp\left(\log(2n) - 2\varepsilon^2 k\right) \Rightarrow$$

$$\forall k > \frac{\log(2n)}{2\varepsilon^2}, \ \mathbb{P}\{\forall j \in [n], |\bar{Y}(j) - \mathbf{p}^T C_j| \leqslant \varepsilon\} \geqslant 1 - \exp\left(\log(2n) - 2\varepsilon^2 k\right) > 0$$

Observe that $\bar{Y}(j)$ is nothing more than the outcome of the inner product of C_j with some $k-$uniform probability distribution, $\hat{\mathbf{p}} \in \Delta_m$, that is uniquely determined by the random variables $X_1, \ldots, X_k$:

$$\forall j \in [n], \ \bar{Y}(j) = \frac{1}{k}\sum_{t\in[k]}Y_t(j) = \frac{1}{k}\sum_{t\in[k]}C_{X_t,j}$$

$$= \sum_{t\in[k]}\sum_{i\in[m]}\left(\frac{1}{k}\cdot\mathbb{I}_{\{X_t=i\}}\right)\cdot C_{i,j} = \sum_{i\in[m]}\underbrace{\left(\frac{1}{k}\sum_{t\in[k]}\mathbb{I}_{\{X_t=i\}}\right)}_{=\hat{p}_i}\cdot C_{i,j}$$

$$= \sum_{i\in[m]}\hat{p}_i\cdot C_{i,j} = \hat{\mathbf{p}}^T C_j$$

So, we conclude that for the probability vector $\hat{\mathbf{p}} = \left(\frac{\sum_{t\in[k]}\cdot\mathbb{I}_{\{X_t=i\}}}{k}\right)_{i\in[m]} \in \Delta_m(k)$ that is produced by the independent random variables $X_1, \ldots, X_k$, it holds that:

$$\forall\varepsilon > \sqrt{\frac{\log(2n)}{2k}}, \ \mathbb{P}\{\forall j \in [n], |\hat{\mathbf{p}}^T C_j - \mathbf{p}^T C_j| \leq \varepsilon\} > 0$$

That is, there is at least one $k-$uniform probability distribution $\hat{\mathbf{p}}$, that certainly has this property.

The following simple observation will be quite useful in our discussion:

Proposition 1 *For any $C \in [0,1]^{m\times n}$ and any probability $\mathbf{p} \in \Delta_m$, the empirical distribution $\hat{\mathbf{p}} \in \Delta_m$ produced by the Approximation Lemma assigns positive probabilities only to rows whose indices belong to $supp(\mathbf{p})$.*

Proof. The result of Althöfer assumes a hypothetical repeated sampling of rows according to the probability vector $\mathbf{p}$. Then it constructs an empirical distribution $\hat{\mathbf{p}}$ by assigning to each row probability proportional to the number of hits of this row in the random experiment. It is proved that the resulting empirical distribution $\hat{\mathbf{p}}$ is actually *with positive probability* a good approximation of $\mathbf{p}$ wrt the columns of the considered matrix C, so long as the number of samples is sufficiently large (this only depends on the number n of columns that have to be taken into account, but not on the dimension of the probability space). This implies that at least one realization of this empirical distribution produced by the random experiment, has this property with certainty. Exactly this repeated sampling procedure also assures that the support of the empirical distribution $\hat{\mathbf{p}}$ will certainly be a subset of the support of the actual distribution $\mathbf{p}$.

We now demonstrate how the Approximation Lemma, along with the previous observation, guarantees the existence of a uniform (2ε)–SuppNE with support sizes at most $\left\lceil \frac{\log(2n)}{2\varepsilon^2} \right\rceil$, for any *constant* $\varepsilon > 0$:

Theorem 2. *Fix any positive constant $\varepsilon > 0$ and an arbitrary $[0,1]$–bimatrix game $\langle A, B \rangle$. There is at least one (k, ℓ)–uniform profile which is also a (2ε)–SuppNE for this game, where $k \le \left\lceil \frac{\log(2n)}{2\varepsilon^2} \right\rceil$ and $\ell \le \left\lceil \frac{\log(2m)}{2\varepsilon^2} \right\rceil$.*

Proof. Assume any profile $(\mathbf{p}, \mathbf{q}) \in NE(A, B)$, which we of course know to exist for any finite game in normal form [23]. We use the Approximation Lemma to assure the existence of some k–uniform strategy $\hat{\mathbf{p}} \in \Delta_m$ with $|supp(\hat{\mathbf{p}})| \le k \equiv \left\lceil \frac{\log(2n)}{2\varepsilon^2} \right\rceil$, such that $|\mathbf{p}^T B_j - \hat{\mathbf{p}}^T B_j| \le \varepsilon, \ \forall j \in [n]$. Similarly, we assume the existence of some ℓ–uniform strategy $\hat{\mathbf{q}} \in \Delta_n$ with $|supp(\hat{\mathbf{q}})| \le \ell \equiv \left\lceil \frac{\log(2m)}{2\varepsilon^2} \right\rceil$, such that $|A^i \mathbf{q} - A^i \hat{\mathbf{q}}| \le \varepsilon, \ \forall i \in [m]$.

Observe now that, trivially, $\hat{\mathbf{p}}^T B - \mathbf{1}^T \cdot \varepsilon \le \mathbf{p}^T B \le \hat{\mathbf{p}}^T B + \mathbf{1}^T \cdot \varepsilon$. Similarly, $A \cdot \hat{\mathbf{q}} - \mathbf{1} \cdot \varepsilon \le A \cdot \mathbf{q} \le A \cdot \hat{\mathbf{q}} + \mathbf{1} \cdot \varepsilon$. Therefore (also exploiting the Nash Property of $(\mathbf{p}, \mathbf{q})$ and the fact that $supp(\hat{\mathbf{p}}) \subseteq supp(\mathbf{p})$) we have:

$$
\begin{array}{llll}
\forall i \in [m], \hat{p}_i > 0 & \overset{/* \text{ Sampling } */}{\Longrightarrow} & p_i > 0 & \\[2mm]
& \overset{/* \text{ Nash Prop. } */}{\Longrightarrow} & A^i \mathbf{q} \ge A^r \mathbf{q}, \ \forall r \in [m] & \\[2mm]
& \overset{/* \text{ Approx. Lemma } */}{\Longrightarrow} & A^i \hat{\mathbf{q}} + \varepsilon \ge A^r \hat{\mathbf{q}} - \varepsilon, \ \forall r \in [m] & \\[2mm]
& \Longrightarrow & A^i \hat{\mathbf{q}} \ge A^r \hat{\mathbf{q}} - 2\varepsilon, \ \forall r \in [m] &
\end{array}
$$

The argument for the column player is identical. Therefore, we conclude that $(\hat{\mathbf{p}}, \hat{\mathbf{q}})$ is a (k, ℓ)–uniform profile that is also a (2ε)–SuppNE for $\langle A, B \rangle$.

3   The Effect of Statistical Conflict in Network Congestion Games

In the present section we focus on the impact of statistical conflict on the quality of a network congestion game, by means of the *price of anarchy* quality measure introduced in [20]. We initially present the model and then we give an upper bound on the blow-up of the price of anarchy, due to the consideration of probability vectors as (mixed) strategies of the players.

3.1   The Network Congestion Game Model

Consider having a set of resources E in a system. For each $e \in E$, let $d_e(\cdot)$ be the **delay** per player that requests his service, as a function of the total usage (ie, the *congestion*) of this resource by all the players. Each such function is considered to be *non-decreasing* in the total usage of the corresponding resource. Each resource may be represented by a pair of points: an entry point to the resource and an exit point from it. So, we represent each resource by an arc from its entry point to its exit point and we associate with this arc the **charging cost** (eg, the delay as a function of the load of this resource) that each player has to pay if he is served by this resource. The entry/exit points of the resources need not be unique; they may coincide in order to express the possibility of offering a *joint service* to players, that consists of a sequence of resources. We denote by V the set of all entry/exit points of the resources in the system. Any nonempty collection of resources corresponding to a directed path in $G \equiv (V, E)$ comprises an **action** in the system.

Let $N \equiv [n]^1$ be the set of players, each willing to adopt some action in the system. $\forall i \in N$, let w_i denote player i's **traffic demand** (eg, the flow rate from a source node to a destination node), while $\mathcal{P}^i \equiv \{a_1^i, \ldots, a_{m_i}^i\} \subseteq 2^E \setminus \emptyset$ (for some $m_i \geq 2$) is the collection of actions, any of which would satisfy player i (eg, alternative routes from a source to a destination node, if G represents a communication network). The collection $\mathcal{P}^i$ is called the *action set* of player i and each of its elements contains at least one resource. Any n−tuple $\varpi \in \mathcal{P} \equiv \times_{i=1}^n \mathcal{P}^i$ is a **pure strategies profile**, or a **configuration** of the players. Any real vector $\mathbf{p} = (\mathbf{p^1}, \mathbf{p^2}, \ldots, \mathbf{p^n})$ s.t. $\forall i \in N$, $\mathbf{p^i} \in \Delta(\mathcal{P}^i) \equiv \{\mathbf{z} \in [0,1]^{m_i} : \sum_{k=1}^{m_i} z_k = 1\}$ is a probability distribution over the set of allowable actions for player i, is called a **mixed strategies profile** for the n players.

A **congestion model** $((\mathcal{P}^i)_{i \in N}, (d_e)_{e \in E})$ typically deals with players of identical demands, and thus the resource delay functions depend only on the *number* of players adopting each action ([11,22,25]). In the more general case, ie, a **weighted congestion model** is the tuple $((w_i)_{i \in N}, (\mathcal{P}^i)_{i \in N}, (d_e)_{e \in E})$. That is, we allow the players to have different (but fixed) demands for service (denoted by their weights) from the whole system, and thus affect the resource delay functions in a different way, depending on their own weights. We denote by $W_{\text{tot}} \equiv \sum_{i \in N} w_i$ and $w_{\max} \equiv \max_{i \in N}\{w_i\}$.

The **weighted congestion game** $\Gamma \equiv (N, E, (w_i)_{i \in N}, (\mathcal{P}^i)_{i \in N}, (d_e)_{e \in E})$ associated with this model, is the game in strategic form with the set of players N and players' demands $(w_i)_{i \in N}$, the set of shared resources E, the action sets $(\mathcal{P}^i)_{i \in N}$ and players' cost functions $(\lambda_{\varpi^i}^i)_{i \in N, \varpi^i \in \mathcal{P}^i}$ defined as follows: For any configuration $\varpi \in \mathcal{P}$ and $\forall e \in E$, let $\Lambda_e(\varpi) = \{i \in N : e \in \varpi^i\}$ be the set of players wishing to exploit resource e according to ϖ (called the **view** of resource e wrt configuration ϖ). We also denote by $x_e(\varpi) \equiv |\Lambda_e(\varpi)|$ the *number* of players using resource e wrt ϖ, whereas $\theta_e(\varpi) \equiv \sum_{i \in \Lambda_e(\varpi)} w_i$ is the **load** of e wrt to ϖ. The **cost** $\lambda^i(\varpi)$ **of player** i **for adopting strategy** $\varpi^i \in \mathcal{P}^i$

---

[1] $\forall k \in \mathbb{N}$, $[k] \equiv \{1, 2, \ldots, k\}$.

in a given configuration ϖ is equal to the cumulative delay $\lambda_{\varpi^i}(\varpi)$ of all the resources comprising this action:

$$\lambda^i(\varpi) = \lambda_{\varpi^i}(\varpi) = \sum_{e \in \varpi^i} d_e(\theta_e(\varpi)). \qquad (1)$$

On the other hand, for a mixed strategies profile $\mathbf{p}$, the **(expected) cost of player i for adopting strategy** $\varpi^i \in \mathcal{P}^i$ wrt $\mathbf{p}$ is

$$\lambda_{\varpi^i}^i(\mathbf{p}) = \sum_{\varpi^{-i} \in \mathcal{P}^{-i}} P(\mathbf{p}^{-i}, \varpi^{-i}) \cdot \sum_{e \in \varpi^i} d_e\left(\theta_e(\varpi^{-i} \oplus \varpi^i)\right) \qquad (2)$$

where, $\varpi^{-i} \in \mathcal{P}^{-i} \equiv \times_{j \neq i} \mathcal{P}^j$ is a configuration of all the players except for i, $\mathbf{p}^{-i} \in \times_{j \neq i} \Delta(\mathcal{P}^j)$ is the mixed strategies profile of all players except for i, $\varpi^{-i} \oplus a$ is the new configuration with i definitely choosing the action $a \in \mathcal{P}^i$, and $P(\mathbf{p}^{-i}, \varpi^{-i}) \equiv \prod_{j \neq i} p_{\varpi^j}^j$ is the occurrence probability of ϖ^{-i} according to $\mathbf{p}^{-i}$.

Remark: We abuse notation a little bit and consider the player costs $\lambda_{\varpi^i}^i$ as functions whose exact definition depends on the other players' strategies: In the general case of a mixed strategies profile $\mathbf{p}$, Eq. (2) is valid and expresses the expected cost of player i wrt $\mathbf{p}$, conditioned on the event that i chooses path ϖ^i. If the other players adopt a pure strategies profile ϖ^{-i}, we get the special form of Eq. (1) that expresses the exact cost of player i choosing action ϖ^i.

A congestion game in which all players are indistinguishable (ie, they have the traffic demands and the same action set), is called **symmetric**. When each player's action set $\mathcal{P}^i$ consists of sets of resources that comprise (simple) paths between a unique origin-destination pair of nodes (s_i, t_i) in (V, E), we refer to a **(multi–commodity) network congestion game**. If additionally all origin-destination pairs of the players coincide with a unique pair (s, t) we have a **single–commodity network congestion game** and then all players share exactly the same action set. Observe that in general a single–commodity network congestion game is not necessarily symmetric because the players may have different demands and thus their cost functions will also differ.

Dealing with Selfish behavior. Fix an arbitrary (mixed in general) strategies profile $\mathbf{p}$ for a congestion game that is described by the tuple $\left((w_i)_{i \in N}, (\mathcal{P}^i)_{i \in N}, (d_e)_{e \in E}\right)$. We say that $\mathbf{p}$ is a **Nash Equilibrium (NE)** if and only if

$$\forall i \in N, \ \forall \alpha, \beta \in \mathcal{P}^i, \ p_\alpha^i > 0 \Rightarrow \lambda_\alpha^i(\mathbf{p}) \leq \lambda_\beta^i(\mathbf{p}).$$

A configuration $\varpi \in \mathcal{P}$ is a **Pure Nash Equilibrium (PNE)** if and only if

$$\forall i \in N, \forall \alpha \in \mathcal{P}^i, \ \lambda^i(\varpi) = \lambda_{\varpi^i}(\varpi) \leq \lambda_\alpha(\varpi^{-i} \oplus \alpha) = \lambda^i(\varpi^{-i} \oplus \alpha).$$

The **social cost** $\mathsf{SC}(\mathbf{p})$ in this congestion game is

$$\mathsf{SC}(\mathbf{p}) = \sum_{\varpi \in \mathcal{P}} P(\mathbf{p}, \varpi) \cdot \max_{i \in N}\{\lambda_{\varpi^i}(\varpi)\} \qquad (3)$$

where $P(\mathbf{p}, \varpi) \equiv \prod_{i=1}^{n} p_{\varpi^i}^i$ is the probability of configuration ϖ occurring, wrt the mixed strategies profile $\mathbf{p}$. The **social optimum** of this game is defined as

$$\mathsf{OPT} = \min_{\varpi \in \mathcal{P}} \left\{ \max_{i \in N} [\lambda_{\varpi^i}(\varpi)] \right\} \tag{4}$$

The **price of anarchy** for this game is then defined as

$$\mathsf{PoA} = \max_{\mathbf{p} \text{ is a NE}} \left\{ \frac{\mathsf{SC}(\mathbf{p})}{\mathsf{OPT}} \right\} \tag{5}$$

Layered Networks. We consider a special family of networks whose behavior wrt the price of anarchy, as we shall see, is asymptotically equivalent to that of the parallel links model of [20] (which is actually a 1-layered network): Let $\ell \geq 1$ be an integer. A directed network $G = (V, E)$ with a distinguished source–destination pair (s, t), $s, t \in V$, is an ℓ-**layered** **network** if every (simple) directed $s - t$ path has length exactly ℓ and each node lies on a directed $s - t$ path. In a layered network there are no directed cycles and all directed paths are simple. In the following, we always use $m = |E|$ to denote the number of edges in an ℓ-layered network $G = (V, E)$.

3.2   Price of Anarchy in Weighted Network Congestion Games

In this subsection we focus our interest on weighted ℓ-layered network congestion games where the resource delays are identical to their loads. Our source for this section is [12]. This case comprises a highly non-trivial generalization of the well–known model of selfish routing of atomic (ie, indivisible) traffic demands via identical parallel channels [20]. The main reason why we focus on this specific category of resource delays is that there exist instances of (even unweighted) congestion games on layered networks that have unbounded price of anarchy even if we only allow linear resource delays. Eg, [26, p. 256] is an example where the price of anarchy is indeed unbounded. This example is easily converted into an ℓ-layered network. The resource delay functions used are either constant, or M/M/1-like (ie, of the form $\frac{1}{c-x}$) delay functions. However, we can be equally bad even in layered networks with *linear* resource delay functions. Such an example is given in [12].

In the following, we restrict our attention to ℓ-layered networks whose resource delays are *equal* to their loads. Our main tool is to interpret a strategies profile as a flow in the underlying network.

Flows and Mixed Strategies Profiles. Fix an arbitrary ℓ-layered network $G = (V, E)$ and a set $N = [n]$ of distinct players willing to satisfy their own traffic demands from the unique source $s \in V$ to the unique destination $t \in V$. Again, $\mathbf{w} = (w_i)_{i \in [n]}$ denotes the varying demands of the players. Fix an arbitrary mixed strategies profile $\mathbf{p} = (p_1, p_2, \ldots, p_n)$ where, for sake of simplicity, we consider that $\forall i \in [n], p_i : \mathcal{P}_{s-t} \mapsto [0, 1]$ is a real function (rather than a vector) assigning non-negative probabilities to the $s-t$ paths of G (which are the allowable actions for player i).

A **feasible flow** for the n players is a function $\rho : \mathcal{P}_{s-t} \mapsto \mathbb{R}_{\geq 0}$ mapping amounts of non-negative traffic (on behalf of all the players) to the $s-t$ paths of G, in such a way that $\sum_{\pi \in \mathcal{P}_{s-t}} \rho(\pi) = W_{\text{tot}} \equiv \sum_{i \in [n]} w_i$. That is, all players' demands are actually satisfied. We distinguish between unsplittable and splittable (feasible) flows. A feasible flow is **unsplittable** if each player's traffic demand is satisfied by a unique path of $\mathcal{P}_{s-t}$. In the general case, any feasible flow is **splittable**, in the sense that the traffic demand of each player is possibly routed over several paths of $\mathcal{P}_{s-t}$.

We map the mixed strategies profile $\mathbf{p}$ to a feasible flow $\rho_{\mathbf{p}}$ as follows: For each $s - t$ path $\pi \in \mathcal{P}_{s-t}$, $\rho_{\mathbf{p}}(\pi) \equiv \sum_{i \in [n]} w_i \cdot p_i(\pi)$. That is, we handle the *expected load traveling along π according to* $\mathbf{p}$ as a splittable flow, where player i routes a fraction of $p_i(\pi)$ of his total demand w_i along π. Observe that, if $\mathbf{p}$ is actually a pure strategies profile, the corresponding flow is then unsplittable. Recall now that for each edge $e \in E$,

$$\theta_e(\mathbf{p}) = \sum_{i=1}^{n} \sum_{\pi : e \in \pi} w_i p_i(\pi) = \sum_{\pi : e \in \pi} \rho_{\mathbf{p}}(\pi) \equiv \theta_e(\rho_{\mathbf{p}})$$

denotes the expected load (and in our case, also the expected delay) of e wrt $\mathbf{p}$, and can be expressed either as a function $\theta_e(\mathbf{p})$ of the mixed profile $\mathbf{p}$, or as a function $\theta_e(\rho_{\mathbf{p}})$ of its associated feasible flow $\rho_{\mathbf{p}}$. As for the expected delay along a path $\pi \in \mathcal{P}_{s-t}$ according to $\mathbf{p}$, this is

$$\theta_\pi(\mathbf{p}) = \sum_{e \in \pi} \theta_e(\mathbf{p}) = \sum_{e \in \pi} \sum_{\pi' \ni e} \rho_{\mathbf{p}}(\pi') = \sum_{\pi' \in \mathcal{P}_{s-t}} |\pi \cap \pi'| \rho_{\mathbf{p}}(\pi') \equiv \theta_\pi(\rho_{\mathbf{p}}) .$$

Let $\theta^{\min}(\rho) \equiv \min_{\pi \in \mathcal{P}_{s-t}} \{\theta_\pi(\rho)\}$ be the minimum expected delay among all $s - t$ paths. From now on for simplicity we drop the subscript of $\mathbf{p}$ from its corresponding flow $\rho_{\mathbf{p}}$, when this is clear by the context. When we compare network flows, two typical measures are those of total latency and maximum latency. For a feasible flow ρ the **maximum latency** is defined as

$$L(\rho) \equiv \max_{\pi : \rho(\pi) > 0} \{\theta_\pi(\rho)\} = \max_{\pi : \exists i, \ p_i(\pi) > 0} \{\theta_\pi(\mathbf{p})\} \equiv L(\mathbf{p}) \tag{6}$$

$L(\rho)$ is nothing but the *maximum expected delay paid by the players*, wrt $\mathbf{p}$. From now on, we use ρ^* and ρ_f^* to denote the optimal unsplittable and splittable flows respectively. The objective of **total latency** is defined as follows:

$$C(\rho) \equiv \sum_{\pi \in \mathcal{P}} \rho(\pi) \theta_\pi(\rho) = \sum_{e \in E} \theta_e^2(\rho) = \sum_{e \in E} \theta_e^2(\mathbf{p}) \equiv C(\mathbf{p}) \tag{7}$$

The second equality is obtained by summing over the edges of π and reversing the order of the summation. We have no direct interpretation of the total latency of a flow to the corresponding mixed profile. Nevertheless, observe that $C(\mathbf{p})$ was used as the **b-potential** function of the corresponding game that proves the existence of a PNE.

Flows at Nash Equilibrium. Let $\mathbf{p}$ be a mixed strategies profile and let ρ be the corresponding flow. For an ℓ-layered network with resource delays equal to the loads, the cost of player i on path π is $\lambda_\pi^i(\mathbf{p}) = \ell w_i + \theta_\pi^{-i}(\mathbf{p})$, where $\theta_\pi^{-i}(\mathbf{p})$ is the expected delay along path π if the demand of player i was removed from the system:

$$\theta_\pi^{-i}(\mathbf{p}) = \sum_{\pi' \in \mathcal{P}} |\pi \cap \pi'| \sum_{j \neq i} w_j p_j(\pi') = \theta_\pi(\mathbf{p}) - w_i \sum_{\pi' \in \mathcal{P}} |\pi \cap \pi'| p_i(\pi') \qquad (8)$$

Thus, $\lambda_\pi^i(\mathbf{p}) = \theta_\pi(\mathbf{p}) + \left[\ell - \sum_{\pi' \in \mathcal{P}} |\pi \cap \pi'| p_i(\pi')\right] w_i$. Observe now that, if $\mathbf{p}$ is a NE, then $L(\mathbf{p}) = L(\rho) \leq \theta^{\min}(\rho) + \ell\, w_{\max}$. Otherwise, the players routing their traffic on a path of expected delay greater than $\theta^{\min}(\rho) + \ell\, w_{\max}$ could improve their delay by defecting to a path of expected delay $\theta^{\min}(\rho)$. We sometimes say that a flow ρ corresponding to a mixed strategies profile $\mathbf{p}$ is a NE with the understanding that it is actually $\mathbf{p}$ which is a NE.

Maximum Latency Versus Total Latency. We show that if the resource delays are equal to their loads, a splittable flow is optimal wrt the objective of maximum latency if and only if it is optimal wrt the objective of total latency. As a corollary, we obtain that the optimal splittable flow defines a NE where all players adopt the same mixed strategy.

Lemma 1 ([12]). *There is a unique feasible flow ρ which minimizes both $L(\rho)$ and $C(\rho)$.*

Proof. For every feasible flow ρ, the average path latency $\frac{1}{W_{\text{tot}}} C(\rho)$ of ρ cannot exceed its maximum latency among the used paths $L(\rho)$:

$$C(\rho) = \sum_{\pi \in \mathcal{P}} \rho(\pi) \theta_\pi(\rho) = \sum_{\pi : \rho(\pi) > 0} \rho(\pi) \theta_\pi(\rho) \leq L(\rho)\, W_{\text{tot}} \qquad (9)$$

A feasible flow ρ minimizes $C(\rho)$ if and only if for every $\pi_1, \pi_2 \in \mathcal{P}$ with $\rho(\pi_1) > 0$, $\theta_{\pi_1}(\rho) \leq \theta_{\pi_2}(\rho)$ (e.g., [4], [24, Section 7.2], [26, Corollary 4.2]). Hence, if ρ is optimal wrt the objective of total latency, for all paths $\pi \in \mathcal{P}, \theta_\pi(\rho) \geq L(\rho)$. Moreover, if $\rho(\pi) > 0$, then $\theta_\pi(\rho) = L(\rho)$. Therefore, if ρ minimizes $C(\rho)$, then the average latency is indeed *equal* to the maximum latency:

$$C(\rho) = \sum_{\pi \in \mathcal{P} : \rho(\pi) > 0} \rho(\pi) \theta_\pi(\rho) = L(\rho) W_{\text{tot}} \qquad (10)$$

Let ρ be the feasible flow that minimizes the total latency and let ρ' be the feasible flow that minimizes the maximum latency. We prove the lemma by establishing that the two flows are identical. Observe that $L(\rho') \geq \frac{C(\rho')}{W_{\text{tot}}} \geq \frac{C(\rho)}{W_{\text{tot}}} = L(\rho)$. The first inequality follows from Ineq. (9), the second from the assumption that ρ minimizes the total latency and the last equality from Eq. (10). On the other hand, it must be $L(\rho') \leq L(\rho)$ because of the assumption that the flow ρ' minimizes the maximum latency. Hence, it must be $L(\rho') = L(\rho)$ and

$C(\rho') = C(\rho)$. In addition, since the function $C(\rho)$ is strictly convex and the set of feasible flows forms a convex polytope, there is a unique flow which minimizes the total latency. Thus, ρ and ρ' must be identical.

The following corollary is an immediate consequence of Lemma 1 and the characterization of the flow minimizing the total latency.

Corollary 1. *A flow ρ minimizes the maximum latency if and only if for every $\pi_1, \pi_2 \in \mathcal{P}$ with $\rho(\pi_1) > 0$, $\theta_{\pi_1}(\rho) \leq \theta_{\pi_2}(\rho)$.*

Proof. By Lemma 1, the flow ρ minimizes the maximum latency if and only if it minimizes the total latency. Then, the corollary follows from the the the fact that ρ minimizes the total latency if and only if for every $\pi_1, \pi_2 \in \mathcal{P}$ with $\rho(\pi_1) > 0$, $\theta_{\pi_1}(\rho) \leq \theta_{\pi_2}(\rho)$ (eg, [24, Section 7.2], [26, Corollary 4.2]).

The following corollary states that the optimal splittable flow defines a mixed NE where all players adopt exactly the same strategy.

Corollary 2. *Let ρ_f^* be the optimal splittable flow and let $\mathbf{p}$ be the mixed strategies profile where every player routes his traffic on each path π with probability $\rho_f^*(\pi)/W_{\text{tot}}$. Then, $\mathbf{p}$ is a NE.*

Proof. By construction, the expected path loads corresponding to $\mathbf{p}$ are equal to the values of ρ_f^* on these paths. Since all players follow exactly the same strategy and route their demand on each path π with probability ρ_f^*/W_{tot}, for each player i,

$$\theta_\pi^{-i}(\mathbf{p}) = \theta_\pi(\mathbf{p}) - w_i \sum_{\pi' \in \mathcal{P}} |\pi \cap \pi'| \frac{\rho_f^*(\pi')}{W_{\text{tot}}} = \left(1 - \frac{w_i}{W_{\text{tot}}}\right)\theta_\pi(\mathbf{p})$$

Since the flow ρ_f^* also minimizes the total latency, for every $\pi_1, \pi_2 \in \mathcal{P}$ with $\rho_f^*(\pi_1) > 0$, $\theta_{\pi_1}(\mathbf{p}) \leq \theta_{\pi_2}(\mathbf{p})$ (eg, [4], [24, Section 7.2], [26, Corollary 4.2]), which also implies that $\theta_{\pi_1}^{-i}(\mathbf{p}) \leq \theta_{\pi_2}^{-i}(\mathbf{p})$. Therefore, for every player i and every $\pi_1, \pi_2 \in \mathcal{P}$ such that player i routes his traffic demand on π_1 with positive probability, $\lambda_{\pi_1}^i(\mathbf{p}) = \ell w_i + \theta_{\pi_1}^{-i}(\mathbf{p}) \leq \ell w_i + \theta_{\pi_2}^{-i}(\mathbf{p}) = \lambda_{\pi_2}^i(\mathbf{p})$. Consequently, $\mathbf{p}$ is a NE.

An Upper Bound on the Social Cost. Next we derive an upper bound on the social cost of every strategy profile whose maximum expected delay (ie, the maximum latency of its associated flow) is within a constant factor from the maximum latency of the optimal unsplittable flow.

Lemma 2. *Let ρ^* be the optimal unsplittable flow, and let $\mathbf{p}$ be a mixed strategies profile and ρ its corresponding flow. If $L(\mathbf{p}) = L(\rho) \leq \alpha L(\rho^*)$, for some $\alpha \geq 1$, then*

$$\mathsf{SC}(\mathbf{p}) \leq 2\,e\,(\alpha + 1)\left(\frac{\log m}{\log \log m} + 1\right)L(\rho^*),$$

where $m = |E|$ denotes the number of edges in the network.

Proof. For each edge $e \in E$ and each player i, let $X_{e,i}$ be the random variable describing the actual load routed through e by i. The random variable $X_{e,i}$ is equal to w_i if i routes his demand on a path π including e and 0 otherwise. Consequently, the expectation of $X_{e,i}$ is equal to $\mathbb{E}\{X_{e,i}\} = \sum_{\pi:e\in\pi} w_i p_i(\pi)$. Since each player selects his path independently, for every fixed edge e, the random variables in $\{X_{e,i}\}_{i\in[n]}$ are independent from each other.

For each edge $e \in E$, let $X_e = \sum_{i=1}^n X_{e,i}$ be the random variable that describes the actual load routed through e, and thus, also the actual delay paid by any player traversing e. X_e is the sum of n independent random variables with values in $[0, w_{max}]$. By linearity of expectation,

$$\mathbb{E}\{X_e\} = \sum_{i=1}^n \mathbb{E}\{X_{e,i}\} = \sum_{i=1}^n w_i \sum_{\pi\ni e} p_i(\pi) = \theta_e(\rho).$$

By applying the standard Hoeffding bound[2] with $w = w_{max}$ and $t = e\kappa \max\{\theta_e(\rho), w_{max}\}$, we obtain that for every $\kappa \geq 1$,

$$\mathbb{P}\{X_e \geq e\kappa \max\{\theta_e(\rho), w_{max}\}\} \leq \kappa^{-e\kappa}.$$

For $m \equiv |E|$, by applying the union bound we conclude that

$$\mathbb{P}\{\exists e \in E : X_e \geq e\kappa \max\{\theta_e(\rho), w_{max}\}\} \leq m\kappa^{-e\kappa} \tag{11}$$

For each path $\pi \in \mathcal{P}$ with $\rho(\pi) > 0$, we define the random variable $X_\pi = \sum_{e\in\pi} X_e$ describing the actual delay along π. The social cost of $\mathbf{p}$, which is equal to the expected maximum delay experienced by some player, cannot exceed the expected maximum delay among paths π with $\rho(\pi) > 0$. Formally,

$$\mathsf{SC}(\mathbf{p}) \leq \mathbb{E}\left\{\max_{\pi:\rho(\pi)>0}\{X_\pi\}\right\}.$$

If for all $e \in E$, $X_e \leq e\kappa \max\{\theta_e(\rho), w_{max}\}$, then for every path $\pi \in \mathcal{P}$ with $\rho(\pi) > 0$,

$$\begin{aligned}
X_\pi = \sum_{e\in\pi} X_e &\leq e\kappa \sum_{e\in\pi} \max\{\theta_e(\rho), w_{max}\} \\
&\leq e\kappa \sum_{e\in\pi}(\theta_e(\rho) + w_{max}) \\
&= e\kappa\left(\theta_\pi(\rho) + \ell w_{max}\right) \\
&\leq e\kappa\left(L(\rho) + \ell w_{max}\right) \\
&\leq e(\alpha + 1)\kappa L(\rho^*)
\end{aligned}$$

---

[2] We use the standard version of Hoeffding bound ([15]): Let $X_1, X_2, \ldots, X_n$ be independent random variables with values in the interval $[0, w]$. Let $X = \sum_{i=1}^n X_i$ and let $\mathbb{E}\{X\}$ denote its expectation. Then, $\forall t > 0$, $\mathbb{P}\{X \geq t\} \leq \left(\frac{e\mathbb{E}\{X\}}{t}\right)^{t/w}$.

The third equality follows from $\theta_\pi(\rho) = \sum_{e \in \pi} \theta_e(\rho)$, the fourth inequality from $\theta_\pi(\rho) \leq L(\rho)$ since $\rho(\pi) > 0$, and the last inequality from the hypothesis that $L(\rho) \leq \alpha L(\rho^*)$ and the fact that $\ell w_{\max} \leq L(\rho^*)$ because ρ^* is an unsplittable flow. Therefore, using Ineq. (11), we conclude that

$$\mathbb{P}\left\{ \max_{\pi:\rho(\pi)>0}\{X_\pi\} \geq \mathrm{e}\,(\alpha+1)\kappa\,L(\rho^*) \right\} \leq m\kappa^{-\mathrm{e}\,\kappa}.$$

In other words, the probability that the actual maximum delay caused by $\mathbf{p}$ exceeds the optimal maximum delay by a factor greater than $2\,\mathrm{e}\,(\alpha+1)\kappa$ is at most $m\kappa^{-\mathrm{e}\,\kappa}$. Therefore, for every $\kappa_0 \geq 2$,

$$\mathsf{SC}(\mathbf{p}) \leq \mathbb{E}\left\{ \max_{\pi:\rho(\pi)>0}\{X_\pi\} \right\} \leq \mathrm{e}\,(\alpha+1)L(\rho^*)\left(\kappa_0 + \sum_{k=\kappa_0}^{\infty} kmk^{-\mathrm{e}\,k}\right)$$

$$\leq \mathrm{e}\,(\alpha+1)L(\rho^*)\left(\kappa_0 + 2m\kappa_0^{-\mathrm{e}\,\kappa_0+1}\right).$$

If $\kappa_0 = \frac{2\log m}{\log\log m}$, then $\kappa_0^{-\mathrm{e}\,\kappa_0+1} \leq m^{-1}$, $\forall m \geq 4$. Thus, we conclude that

$$\mathsf{SC}(\mathbf{p}) \leq 2\,\mathrm{e}\,(\alpha+1)\left(\frac{\log m}{\log\log m}+1\right)L(\rho^*).$$

Bounding the Price of Anarchy. Our final step is to show that the maximum expected delay of every NE is a good approximation to the optimal maximum latency. Then, we can apply Lemma 2 to bound the price of anarchy for our selfish routing game.

Lemma 3. *For every flow ρ corresponding to a mixed strategies profile $\mathbf{p}$ at NE, $L(\rho) \leq 3L(\rho^*)$.*

Proof. The proof is based on Dorn's Theorem [9] which establishes strong duality in quadratic programming[3]. We use quadratic programming duality to prove that for any flow ρ at Nash equilibrium, the minimum expected delay $\theta^{\min}(\rho)$ cannot exceed $L(\rho_f^*) + \ell\,w_{\max}$. This implies the lemma because $L(\rho) \leq \theta^{\min}(\rho) + \ell\,w_{\max}$, since ρ is at Nash equilibrium, and $L(\rho^*) \geq \max\{L(\rho_f^*), \ell\,w_{\max}\}$, since ρ^* is an unsplittable flow.

Let Q be the square matrix describing the number of edges shared by each pair of paths. Formally, Q is a $|\mathcal{P}| \times |\mathcal{P}|$ matrix and for every $\pi, \pi' \in \mathcal{P}$, $Q[\pi, \pi'] = |\pi \cap \pi'|$. By definition, Q is symmetric. Next we prove that Q is positive semi-definite[4].

---

[3] Let $\min\{x^T Q x + c^T x : Ax \geq b, x \geq \mathbf{0}\}$ be the primal quadratic program. The Dorn's dual of this program is $\max\{-y^T Q y + b^T u : A^T u - 2Qy \leq c, u \geq \mathbf{0}\}$. Dorn [9] proved strong duality when the matrix Q is symmetric and positive semi-definite. Thus, if Q is symmetric and positive semi-definite and both the primal and the dual programs are feasible, their optimal solutions have the same objective value.

[4] An $n \times n$ matrix Q is positive semi-definite if for every vector $x \in \mathbb{R}^n$, $x^T Q x \geq 0$.

$$x^T Q x = \sum_{\pi \in \mathcal{P}} x(\pi) \sum_{\pi' \in \mathcal{P}} Q[\pi, \pi'] x(\pi')$$

$$= \sum_{\pi \in \mathcal{P}} x(\pi) \sum_{\pi' \in \mathcal{P}} |\pi \bigcap \pi'| x(\pi')$$

$$= \sum_{\pi \in \mathcal{P}} x(\pi) \sum_{e \in \pi} \sum_{\pi' : e \in \pi'} x(\pi')$$

$$= \sum_{\pi \in \mathcal{P}} x(\pi) \sum_{e \in \pi} \theta_e(x)$$

$$= \sum_{e \in E} \theta_e(x) \sum_{\pi : e \in \pi} x(\pi)$$

$$= \sum_{e \in E} \theta_e^2(x) \geq 0$$

First recall that for each edge e, $\theta_e(x) \equiv \sum_{\pi : e \in \pi} x(\pi)$. The third and the fifth equalities follow by reversing the order of summation. In particular, in the third equality, instead of considering the edges shared by π and π', for all $\pi' \in \mathcal{P}$, we consider all the paths π' using each edge $e \in \pi$. On both sides of the fifth inequality, for every edge $e \in E$, $\theta_e(x)$ is multiplied by the sum of $x(\pi)$ over all the paths π using e.

Let ρ also denote the $|\mathcal{P}|$-dimensional vector corresponding to the flow ρ. Then, the π-th coordinate of $Q\rho$ is equal to the expected delay $\theta_\pi(\rho)$ on the path π, and the total latency of ρ is $C(\rho) = \rho^T Q \rho$.

Therefore, the problem of computing a feasible splittable flow of minimum total latency is equivalent to computing the optimal solution to the following quadratic program: $\min\{\rho^T Q \rho : \mathbf{1}^T \rho \geq W_{\text{tot}}, \rho \geq \mathbf{0}\}$, where $\mathbf{1}/\mathbf{0}$ denotes the $|\mathcal{P}|$-dimensional vector having $1/0$ in each coordinate. Also notice that no flow of value strictly greater than W_{tot} can be optimal for this program. This quadratic program is clearly feasible and its optimal solution is ρ_f^* (Lemma 1).

The Dorn's dual of this quadratic program is: $\max\{z W_{\text{tot}} - \rho^T Q \rho : 2Q\rho \geq \mathbf{1}z, z \geq 0\}$ (e.g., [9], [3, Chapter 6]). We observe that any flow ρ can be regarded as a feasible solution to the dual program by setting $z = 2\,\theta^{\min}(\rho)$. Hence, both the primal and the dual programs are feasible. By Dorn's Theorem [9], the objective value of the optimal dual solution is exactly $C(\rho_f^*)$. More specifically, the optimal dual solution is obtained from ρ_f^* by setting $z = 2\theta^{\min}(\rho_f^*)$. Since $L(\rho_f^*) = \theta^{\min}(\rho_f^*)$ and $C(\rho_f^*) = L(\rho_f^*)W_{\text{tot}}$, the objective value of this solution is $2\theta^{\min}(\rho_f^*)W_{\text{tot}} - C(\rho_f^*) = C(\rho_f^*)$.

Let ρ be any feasible flow at Nash equilibrium. Setting $z = 2\,\theta^{\min}(\rho)$, we obtain a dual feasible solution. By the discussion above, the objective value of the feasible dual solution $(\rho, 2\,\theta^{\min}(\rho))$ cannot exceed $C(\rho_f^*)$. In other words,

$$2\,\theta^{\min}(\rho)\,W_{\text{tot}} - C(\rho) \leq C(\rho_f^*) \tag{12}$$

Since ρ is at Nash equilibrium, $L(\rho) \leq \theta^{\min}(\rho) + \ell\, w_{\max}$. In addition, by Ineq. (9), the average latency of ρ cannot exceed its maximum latency. Thus,

$$C(\rho) \le L(\rho) \, W_{\text{tot}} \le \theta^{\min}(\rho) \, W_{\text{tot}} + \ell \, w_{\max} \, W_{\text{tot}}$$

Combining the inequality above with Ineq. (12), we obtain that $\theta^{\min}(\rho) \, W_{\text{tot}} \le C(\rho_f^*) + \ell \, w_{\max} \, W_{\text{tot}}$. Using $C(\rho_f^*) = L(\rho_f^*) \, W_{\text{tot}}$, we conclude that $\theta^{\min}(\rho) \le L(\rho_f^*) + \ell \, w_{\max}$.

The following theorem is an immediate consequence of Lemma 3 and Lemma 2.

Theorem 3 ([12]). *The price of anarchy of any ℓ-layered network congestion game with resource delays equal to their loads, is at most* $8 \, \mathrm{e} \left(\frac{\log m}{\log \log m} + 1 \right)$.

A recent development which is complementary to the last theorem is the following which we state without a proof:

Theorem 4 ([13]). *The price of anarchy of any unweighted, single–commodity network congestion game with resource delays* $(d_e(x) = a_e \cdot x, a_e \ge 0)_{e \in E}$, *is at most* $24 \, \mathrm{e} \left(\frac{\log m}{\log \log m} + 1 \right)$.

3.3   The Pure Price of Anarchy in Congestion Games

In this last subsection we overview some recent advances in the *Pure Price of Anarchy* (PPoA) of congestion games, that is, the worst-case ratio of the social cost of a PNE over the social optimum of the game.

The case of linear resource delays has been extensively studied in the literature. The PPoA wrt the total latency objective has been proved that it is $\frac{3+\sqrt{5}}{2}$, even for weighted multi–commodity network congestion games [2,7]. This result is also extended to the case of mixed equilibria. For the special case of identical players it has been proved (independently by the papers [2,7]) that the PPoA drops down to $5/2$. When considering identical users and single–commodity network congestion games, the PPoA is again $5/2$ wrt the maximum latency objective, but explodes to $\Theta(\sqrt{n})$ for multi–commodity network congestion games ([7]). Earlier it was implicitly proved by [12] that the PPoA of any weighted congestion game on a layered network with resource delays identical to the congestion, is at most 3.

4   The Effect of Coalitions in Parallel Links Congestion Games

4.1   The Coalitional KP Model

We consider a collection $M = [m]$ of identical parallel machines (the resources) and a collection $N = [n]$ of jobs (the users)[5]. Each user may be allocated to any of the m available resources. Associated with each user $i \in [n]$ is an integer **service demand** $w_i \in \mathbb{N}_+$ (eg, the number of elementary operations for serving user i). We denote by $\widetilde{W} = \{w_i\}_{i \in [n]}$ the *multiset* of the users' demands.

---

[5] For any integer $k \ge 1$, $[k] \equiv \{1, \dots, k\}$.

Definition 4 (Coalitions). *A set of* $k \geq 1$ *static coalitions (the players)* $C_1, \ldots, C_k$ *is a fixed partition of* $\widetilde{W}$ *into* k *nonempty multisets: (i)* $\cup_{j \in [k]} C_j = \widetilde{W}$, *(ii)* $C_j \neq \emptyset$, $\forall j \in [k]$, *and (iii)* $C_i \cap C_j = \emptyset$, $\forall i, j \in [k] : i \neq j$.[6]

For $j \in [k]$ let $C_j = \{w_j^1, \ldots, w_j^{n_j}\}$, so that $\sum_{j=1}^{k} n_j = n$. Let $W_j = \sum_{i=1}^{n_j} w_j^i$ be the cumulative service demand required by coalition C_j, let $w_j^{\max} = \max_{i \in [n_j]} \{w_j^i\}$ be the largest demand handled by coalition C_j, and let $W_{\text{tot}} = \sum_{j \in [k]} W_j$ be the overall service demand required by the system. Wlog assume that $w_j^1 \geq \cdots \geq w_j^{n_j}$ for all $j \in [k]$.

Strategies and Profiles. A **pure strategy** $\sigma_j = (\sigma_j^i)_{i \in [n_j]}$ for coalition C_j defines the deterministic selection of a resource $\sigma_j^i \in M$ for each $w_j^i \in C_j$. Denote by $\S_j$ the set of all pure strategies available to coalition C_j. Clearly, $\S_j = M^{n_j}$. We denote by $S_j(\sigma_j) = \{\ell \in [m] : \exists i \in [n_j] \text{ with } \sigma_j^i = \ell\}$ the set of resources serving some user of coalition j in pure strategy σ_j.

A **mixed strategy** for coalition C_j is a probability distribution $\mathbf{p_j}$ on the set $\S_j$ of its pure strategies (ie, a point of the simplex $\Delta(\S_j) \equiv \{\mathbf{q} \in \mathbb{R}^{n_j} : \mathbf{q} \geq \mathbf{0}; \ \mathbf{1}^T \mathbf{q} = 1\}$). In order to indicate the probability of pure strategy σ_j being chosen by C_j when the mixed strategy $\mathbf{p_j}$ has been adopted, we use (for sake of simplicity) the functional notation $p_j(\sigma_j)$, rather than the coordinate of the vector $\mathbf{p_j}$ corresponding to σ_j.

A **pure strategies profile** or **configuration** for the coalitions is a collection $\sigma = (\sigma_j)_{j \in [k]}$ of pure strategies, one per coalition. $\S \equiv \times_{j \in [k]} \S_j$ is the set of all the possible configurations of the game (called the **configuration space**). Let (σ_{-j}, α_j) denote the configuration resulting from a configuration σ when coalition C_j unilaterally changes its pure strategy from σ_j to α_j. Similarly, we call the simplotope $\Delta(\S) \equiv \times_{j \in [k]} \Delta(\S_j)$ the **mixed strategies space** of the coalitional game. A **mixed strategies profile** $\mathbf{p} = (\mathbf{p_j})_{j \in [k]} \in \Delta(\S)$ is a collection of mixed strategies, one per coalition. $(\mathbf{p_{-j}}, \mathbf{q_j})$ is the mixed strategies profile in which all players except for player j adopt the strategies indicated by $\mathbf{p}$, while player j adopts strategy $\mathbf{q_j}$ (rather than $\mathbf{p_j}$).

The **support** of coalition $j \in [k]$ in the mixed profile $\mathbf{p}$ is the set $S_j(\mathbf{p}) = \{\sigma_j \in \S_j : p_j(\sigma_j) > 0\}$; thus $S_j(\mathbf{p})$ is the set of pure strategies that coalition j chooses with non-zero probability. A profile $\mathbf{p}$ having $S_j = \S_j$ for all the coalitions, is called a **fully mixed** profile. A profile $\mathbf{p}$ is said to be a **generalized fully mixed** profile, if for each coalition $j \in [k]$ it holds that it assigns positive probability mass to all, but only those *optimum* m-partitions of its own users' weights to the links, ie, to all those configurations (but only them) $\sigma_j \in \S_j$ which are lexicographically minimum wrt the load vector (for the definition of the load of a link see next paragraph).

---

[6] The union and intersection operations are over *multisets* (and return multisets). An equivalent representation that only uses sets, would be to consider each user as a pair (i, w_i) of the user's (unique) identifier and its own (not necessarily unique) service demand. In this representation the meaning of the union and intersection operations is clear.

A special case of particular interest in scheduling literature, is when the coalitions are enforced to eventually choose *consecutive resources* for their own tasks (or similarly, in a single subinterval in case of interval scheduling). When the coalitions are forced to choose only this kind of pure strategies for their own users, then we shall refer to the **Coalitional Chains** model.

Selfish Costs. Fix a configuration $\sigma = (\sigma_j)_{j \in [k]}$. Define as **load** on resource $\ell \in M$ due to coalition C_j, the cumulative demand induced on ℓ by this coalition: $\theta_\ell(\sigma_j) \equiv \sum_{i \in [n_j]: \sigma_j^i = \ell} w_j^i$. The **total load** on resource $\ell \in M$ is the total demand on ℓ with respect to σ, ie, $\theta_\ell(\sigma) = \sum_{j=1}^k \theta_\ell(\sigma_j)$. Similarly, the load induced on resource $\ell \in M$ by all the coalitions except for coalition C_j, is $\theta_\ell(\sigma_{-j}) = \sum_{r \in [k] \setminus \{j\}} \theta_\ell(\sigma_r)$. The **selfish cost** $\lambda_j(\sigma)$ of coalition C_j is the maximum load over the set of resources it employs for its users: $\lambda_j(\sigma) = \max_{\ell \in S_j(\sigma_j)} \{\theta_\ell(\sigma)\}$.

For a mixed profile **p**, the load on each resource $\ell \in M$ becomes a random variable induced by the probability distributions $\mathbf{p_j}$ for all $j \in [k]$. More precisely, let $\theta_\ell(\mathbf{p_j}) = \sum_{\sigma_j \in \Sigma_j} p_j(\sigma_j) \theta_\ell(\sigma_j)$ be the expected load induced on resource ℓ by coalition j according to mixed strategy $\mathbf{p_j}$. The **expected load** on resource $\ell \in M$, denoted by $\theta_\ell(\mathbf{p})$, is the expectation of the load on ℓ according to **p**. Formally,

$$\theta_\ell(\mathbf{p}) = \sum_{\sigma \in \S} \left[\left(\prod_{j \in [k]} p_j(\sigma_j) \right) \cdot \theta_\ell(\sigma) \right] = \sum_{j \in [k]} \theta_\ell(\mathbf{p_j})$$

We use $\theta_\ell(\mathbf{p_{-j}})$ to denote the expected load that all the coalitions except for coalition C_j induce on resource ℓ. Formally,

$$\theta_\ell(\mathbf{p_{-j}}) = \sum_{\sigma_{-j} \in \S_{-j}} \left[\left(\prod_{r \in [k] \setminus \{j\}} p_r(\sigma_r) \right) \cdot \theta_\ell(\sigma_{-j}) \right] = \sum_{r \in [k] \setminus \{j\}} \theta_\ell(\mathbf{p_r})$$

The **conditional expected selfish cost** of coalition j adopting the pure strategy $\sigma_j \in \S_j$, given that the other coalitions follow the strategies indicated by **p**, is

$$\lambda_j(\mathbf{p_{-j}}, \sigma_j) = \max_{\ell \in S_j(\sigma_j)} \{\theta_\ell(\sigma_j) + \theta_\ell(\mathbf{p_{-j}})\}$$

In words, coalition C_j pays for the conditional expectation of its selfish cost, had it adopted the pure strategy $\sigma_j \in \S_j$. This is because coalition C_j has to encounter all the possible alternatives for serving its own users prior to the other coalitions' determination of their actual action, knowing only their probability distributions. The **expected selfish cost** of coalition C_j is defined as the expectation of coalition C_j's conditional expected cost, over all possible actions that can be taken: $\lambda_j(\mathbf{p}) = \sum_{\sigma_j \in \S_j} [p_j(\sigma_j) \cdot \lambda_j(\mathbf{p_{-j}}, \sigma_j)]$.

Remark: Note that each coalition pays for the *expected maximum load* that it would cause if it was on its own, plus the *expected loads* caused by the other coalitions to each of the resources.

Nash Equilibria. The definition of expected selfish costs completes the definition of the finite normal form game involving the k static coalitions (the players) of users that are to be served by the m shared resources: $\Gamma = \langle [k], (\S_j)_{j \in [k]}, (\lambda_j)_{j \in [k]} \rangle$. We are interested in the induced Nash Equilibria [23] of Γ. Informally, a Nash Equilibrium is a (pure or mixed) profile such that no coalition can reduce its expected selfish cost by unilaterally changing its strategy. Formally:

Definition 5 (Nash Equilibrium). *A pure strategies profile* $\sigma = (\sigma_j)_{j \in [k]}$ *is a* **Pure Nash Equilibrium (PNE)** *for* Γ *if,* $\forall j \in [k]$, $\forall \alpha_j \in \S_j$, $\lambda_j(\sigma) \leq \lambda_j(\sigma_{-j}, \alpha_j)$. *A mixed strategies profile* $\mathbf{p}$ *is a* **Nash Equilibrium (NE)** *if,* $\forall j \in [k]$, $\forall \sigma_j \in \S_j$, *it holds that* $p_j(\sigma_j) > 0 \;\Rightarrow\; \sigma_j \in \arg\min_{\alpha_j \in \S_j} \{\lambda_j(\mathbf{p}_{-j}, \alpha_j)\}$.

Assume now that the players, rather than trying unilateral changes of strategies, also consider joint changes in groups of at most r players. These changes are considered to be selfish if they improve the coalitional cost of the players participating in them.

We call a pure strategies profile which is robust against any sort of selfish coalitional $(\leq r)$–move, an r–**robust** PNE. For example, the traditional PNE when no coalitions are allowed, are in our terminology equivalent to the 1–robust PNE. When no more than pairs of players are allowed to form coalitions, then our stable points are the 2–robust PNE. More formally:

Definition 6 (Robust Nash Equilibrium). *A pure strategies profile* $\sigma = (\sigma_j)_{j \in [k]}$ *is an* r–**robust PNE** *for* Γ *if,* $\forall S \subseteq [n] : 1 \leq |S| \leq r$, $\forall \alpha_S \in [m]^{|S|}$, $\lambda_S(\sigma) \leq \lambda_S(\sigma_{-S}, \alpha_S)$, *where we denote by* $\lambda_S(\sigma)$ *the expected selfish cost of coalition* S *in* σ. *Similarly, a mixed strategies profile* $\mathbf{p}$ *is an* r–**robust NE** *if,* $\forall S \subseteq [n] : 1 \leq |S| \leq r$, $\forall \alpha_S \in [m]^{|S|}$, *it holds that* $p_S(\alpha_S) > 0 \;\Rightarrow\; \alpha_S \in \arg\min_{\alpha_S \in [m]^{|S|}} \{\lambda_S(\mathbf{p}, \alpha_S)\}$.

Social Cost, Social Optimum and Price of Anarchy. For any configuration $\sigma = (\sigma_j)_{j \in [k]}$ we define the **social cost**, denoted $\mathsf{SC}(\sigma)$, to be the maximum load over the set of the shared resources M, with respect to σ. That is, $\mathsf{SC}(\sigma) = \max_{\ell \in M} \{\theta_\ell(\sigma)\} = \max_{j \in [k]} \{\lambda_j(\sigma)\}$. For any mixed profile $\mathbf{p}$ the social cost is defined as the expectation, over all random choices of the coalitions, of the maximum load over the set of resources: $\mathsf{SC}(\mathbf{p}) = \sum_{\sigma \in \S} \left(\prod_{j=1}^{k} p_j(\sigma_j) \right) \cdot \max_{\ell \in M} \{\theta_\ell(\sigma)\}$.

Now let σ^* be a configuration that minimizes the social cost function, ie, $\sigma^* \in \arg\min_\sigma \{\mathsf{SC}(\sigma)\}$. Thus σ^* is an optimal configuration of the set of loads $\widetilde{W}$ to the set of resources M. We denote its value by $\mathsf{OPT} = \mathsf{SC}(\sigma^*)$. The **Price of Anarchy** (also referred to as **Coordination Ratio**) [20], is the worst–case ratio of the social cost paid at any NE, over the value of the social optimum of the game: $\mathsf{PoA} = \max_{\mathbf{p} \text{ is NE}} \left\{ \frac{\mathsf{SC}(\mathbf{p})}{\mathsf{OPT}} \right\}$.

Improvement Paths. When we discuss convergence issues, we shall frequently refer to the notion of improvement paths: These are sequences of configurations for the coalitions (ie, points in §), such that any two consecutive configurations differ only in the pure strategy of *exactly one* coalition, and additionally the cost of this unique coalition is *strictly less* in the latter configuration than in the former one. Observe that an improvement path is not necessarily acyclic, as it would be the case in the graph–theoretic notion of a path. The only demand is that the single coalition that alters its pure strategy (between two consecutive configurations in the path) has a reason to do so. If it is true that *any* improvement path of a normal form game is acyclic, then we say that this game holds the **Finite Improvement Property (FIP)**. This is a very nice property since it is a sufficient (but not necessary) condition for the existence of PNE for the game.

4.2   Price of Anarchy in the Coalitional KP Model

Assume that there is a single coalition $C_1 = \widetilde{W}$, ie $k = 1$. In this case, any NE is an optimum assignment of C_1 to M and vice versa. Hence in any NE σ, $\mathsf{SC}(\sigma) = \mathsf{OPT}$ and thus $\mathsf{R} = 1$. On the other hand, the case where $k = n$ reduces to the standard KP-model [20], for which $\mathsf{PoA} = \Theta\left(\frac{\log m}{\log \log m}\right)$ [19].

We now prove that for every $k \in [n]$, the price of anarchy is $\Theta(\min\{k, \frac{\log m}{\log \log m}\})$. The lower bound (Theorem 5) holds even for identical tasks and coalitions of equal cardinality. The upper bound (Theorem 6) holds for weighted tasks and arbitrary coalitions.

The Lower Bound

Theorem 5. *The price of anarchy is* $\Omega(\min\{k, \frac{\log m}{\log \log m}\})$ *even for identical tasks and coalitions of equal cardinality.*

Proof. We consider m identical parallel links and m unit size tasks partitioned into $k \geq 2$ coalitions each with $r \equiv m/k$ tasks (wlog we assume that m/k is an integer). We say that a coalition $j \in [k]$ **hits** a link $\ell \in [m]$ if j assigns at least one of its tasks to ℓ. We first prove a lower bound for the Coalitional Chains Model.

Lemma 4. *In the Coalitional Chains Model, when the number of coalitions is $k = m^\varepsilon$ for arbitrary constant $\varepsilon \in (0, 1]$, the price of anarchy is* $\mathsf{PoA} = \Omega\left(\frac{\log m}{\log \log m}\right)$.

Proof. It is easy to construct a very simple and natural (especially in scheduling literature) mixed profile that is indeed a NE and achieves the lower bound of $\Omega\left(\frac{\log m}{\log \log m}\right)$, for the case where the number of coalitions is $k = m^\varepsilon$, for arbitrary constant $\varepsilon \in (0, 1]$. This example is the following: Each coalition contains exactly $r = m/k = m^{1-\varepsilon}$ unit size tasks. We consider the following mixed profile **p** for the coalitions: Assuming that the m links form a cycle, each coalition $j \in [k]$ chooses uniformly at random a link $\ell_j \in [m]$ as its starting point, and then assigns its r weights to r consecutive links $\ell_j, \ell_j + 1 (\text{mod } m), \ldots, \ell_j + r - 1 (\text{mod } m)$.

We show first that this is indeed a NE: Observe that for any link $\ell \in [m]$, the expected load wrt any subset of coalitions using the profile $\mathbf{p}$ is the same: $\forall \ell \in [m], \theta_\ell(\mathbf{p}) = \sum_{j \in [k]} 1 \cdot \mathbb{P}\{j \text{ hits link } \ell\} = \sum_{j \in [k]} n^{-\varepsilon} = 1$ and $\forall \ell \in [m], \forall j \in [k], \theta_\ell(\mathbf{p_{-j}}) = \sum_{j' \in [k] \setminus \{j\}} 1 \cdot \mathbb{P}\{j' \text{ hits link } \ell\} = \sum_{j' \in [k] \setminus \{j\}} n^{-\varepsilon} = 1 - n^{-\varepsilon}$. Observe now that, since for any coalition $j \in [k]$ the expected load of the other players is exactly the same and there is no chance (wrt $\mathbf{p}$) that two balls of C_j fall into the same bin, the expected cost of C_j is $\lambda_j(\mathbf{p}) = 1 + 1 - n^{-\varepsilon} = 2 - n^{-\varepsilon}$. On the other hand, $\forall \sigma_j \in \S_j, \lambda_j(\mathbf{p_{-j}}, \sigma_j) = \max_{\ell \in S_j(\sigma_j)}\{\theta_\ell(\sigma_j) + \theta_\ell(\mathbf{p_{-j}})\} \geq 1 + 1 - n^{-\varepsilon} = \lambda_j(\mathbf{p})$, Therefore, we conclude that $\mathbf{p}$ is a mixed NE for the coalitional game.

Next we prove that the coordination ratio for $\mathbf{p}$ is $\Omega\left(\frac{\log m}{\log \log m}\right)$. To see this, simply observe that the expected maximum load induced by $\mathbf{p}$ is lower bounded by the expected maximum load of any subset of links. We call a **chain** any set of r consecutive links in M (modulo m). Each player then chooses a chain independently and uniformly at random to allocate his r weights. Thus, a specific chain hits a link $\ell \in [m]$ if ℓ belongs to this chain. Consider the subset of links $M' \equiv \{1, r+1, 2r+1, \ldots, (k-1)r + 1\}$. The links of this subset have the property that any chain will hit exactly one of the links in M'. Thus, the expected maximum load (wrt to the profile $\mathbf{p}$) among the links of M' equals the expected maximum load that we face when throwing k identical balls (the choices of the players' chains) to k identical bins (the links included in M'). But this is known that it equals $\Theta\left(\frac{\log k}{\log \log k}\right)$. Therefore, the social cost of $\mathbf{p}$ is lower bounded as follows: $\mathsf{SC}(\mathbf{p}) = \Omega\left(\frac{\log k}{\log \log k}\right) = \Omega\left(\frac{\log m^\varepsilon}{\log \log m^\varepsilon}\right) = \Omega\left(\frac{\varepsilon \log m}{\log \varepsilon + \log \log m}\right)$. The social optimum of this example is clearly 1. So, for any constant $1 \geq \varepsilon > 0$, we have that $\mathsf{PoA} = \Omega\left(\frac{\log m}{\log \log m}\right)$. $\qquad\square$

General Case. As before, we consider m identical parallel links and m unit size tasks partitioned into $k \geq 2$ coalitions with $r \equiv m/k$ tasks each. The social optimum is equal to 1. The lower bound is established for the following generalized fully mixed NE: Let $\mathbf{p}$ be the mixed profile where every coalition chooses r links uniformly at random without replacement and assigns a task to every chosen link. In other words, every coalition assigns its tasks to any particular combination of r links with probability $1/\binom{m}{r}$. The probability that coalition j hits link ℓ is $1 - \binom{m-1}{r}/\binom{m}{r} = \frac{r}{m} = \frac{1}{k}$. Every coalition assigns at most one of its tasks to every link. It is simple to show that $\mathbf{p}$ is a NE for the coalitional game (see [14]).

For every link $\ell \in [m]$ and every positive integer $\rho \leq k$, let $\mathcal{E}_\ell(\rho) \in \{0, 1\}$ be the indicator variable of the event that link ℓ is hit by less than ρ coalitions in the outcome produced by $\mathbf{p}$. For convenience, we call a link ℓ **light** if it is hit by less than ρ coalitions, ie if $\mathcal{E}_\ell(\rho) = 1$, and **heavy** otherwise. The number of coalitions hitting a given link ℓ is the sum of k independent 0-1 random variables each becoming 1 with probability $1/k$. It is easy to calculate an upper bound on the probability that a given link is light.

$$\mathbb{P}\left\{\mathcal{E}_\ell(\rho) = 0\right\} \geq \binom{k}{\rho}\left(\frac{1}{k}\right)^\rho\left(1 - \frac{1}{k}\right)^{k-\rho} \geq \frac{1}{\exp\rho^\rho} \Rightarrow$$

$$\mathbb{P}\left\{\mathcal{E}_\ell(\rho) = 1\right\} \leq 1 - \frac{1}{\exp\rho^\rho} \tag{13}$$

where we use that $\binom{k}{\rho} \geq (\frac{k}{\rho})^\rho$ and that for every integer $k \geq 2$, $(1 - \frac{1}{k})^{k-1} \geq \frac{1}{\exp}$.

The crucial observation is that the link lightness events are negatively associated[7]. Therefore, for every positive integer $\rho \leq k$,

$$\mathbb{P}\left\{\mathcal{E}_1(\rho) = \cdots = \mathcal{E}_m(\rho) = 1\right\} \leq \left(1 - \frac{1}{\exp\rho^\rho}\right)^m \leq \exp^{-\frac{m}{\exp\rho^\rho}} \tag{14}$$

The second inequality follows from $e^{-x} \geq 1 - x$. The first inequality is established in Lemma 5 below similarly to [16, Lemma 1].

Before we formally prove (14), we show that it indeed implies the lemma. In particular, we show that unless $\rho = O(\min\{k, \frac{\log m}{\log\log m}\})$, it is almost certain that in the configuration chosen wrt $\mathbf{p}$, at least one of the links is heavy. Indeed for every constant $c \geq 1$, there is a constant $\alpha \in (0, 1)$ such that for all positive integers $\rho \leq \frac{\alpha\log m}{\log\log m}$, $\rho^\rho \leq \frac{m}{c\exp\log m}$ (note that α tends asymptotically to 1). Therefore, $\exp^{-\frac{m}{\exp\rho^\rho}} \leq m^{-c}$. In simple words, with probability at least $1 - m^{-c}$, there is some link hit by at least $\min\{k, \frac{\alpha\log m}{\log\log m}\}$ coalitions. If $k = \Omega(\frac{\log m}{\log\log m})$, the social cost of $\mathbf{p}$ is $\Theta(\frac{\log m}{\log\log m})$. If $k = o(\frac{\log m}{\log\log m})$, the social cost of $\mathbf{p}$ is $\Theta(k)$. Since the optimal social cost is 1, this implies the lower bound of Theorem 5.

The following lemma establishes that the link lightness events are negatively associated.

Lemma 5. *For any fixed positive integer $\rho \leq k$,*

$$\mathbb{P}\left\{\mathcal{E}_1(\rho) = \cdots = \mathcal{E}_m(\rho) = 1\right\} \leq \prod_{\ell=1}^m \mathbb{P}\left\{\mathcal{E}_\ell(\rho) = 1\right\}$$

Proof. For simplicity, we assume an arbitrary fixed threshold $\rho \leq k$ and denote the event that a link ℓ is light wrt ρ simply by $\mathcal{E}_\ell$ (instead of $\mathcal{E}_\ell(\rho)$). We first prove that for every $\ell \in \{2, \ldots, m\}$,

$$\mathbb{P}\left\{\mathcal{E}_1 = \cdots = \mathcal{E}_{\ell-1} = 1 | \mathcal{E}_\ell = 1\right\} \leq \mathbb{P}\left\{\mathcal{E}_1 = \cdots = \mathcal{E}_{\ell-1} = 1 | \mathcal{E}_\ell = 0\right\} \tag{15}$$

For all $l \in [m]$, let θ_l denote the number of coalitions hitting l in the outcome produced by $\mathbf{p}$. We fix an $\ell \in \{2, \ldots, m\}$. To establish (15), we consider the first ℓ links and fix the total number of hits in the links of $[\ell]$. In particular, let M be an arbitrary fixed integer such that $M = \sum_{l\in[\ell]} \theta_l$. We observe that

$$\mathbb{P}\left\{\mathcal{E}_1 = \cdots = \mathcal{E}_{\ell-1} = 1 | \mathcal{E}_\ell = 1; \sum_{l\in[\ell]} \theta_l = M\right\}$$

$$= \mathbb{P}\left\{\mathcal{E}_1 = \cdots = \mathcal{E}_{\ell-1} = 1 | \theta_\ell \leq \rho - 1; \sum_{l\in[\ell]} \theta_l = M\right\}$$

$$= \mathbb{P}\left\{\mathcal{E}_1 = \cdots = \mathcal{E}_{\ell-1} = 1 | \sum_{l\in[\ell-1]} \theta_l \geq M - \rho + 1; \sum_{l\in[\ell]} \theta_l = M\right\} \tag{16}$$

---

[7] The negative association of bin occupancies in the classical "balls and bins" experiment is established eg in [10].

whereas

$$\mathbb{P}\left\{\mathcal{E}_1 = \cdots = \mathcal{E}_{\ell-1} = 1 | \mathcal{E}_\ell = 0; \ \sum_{l \in [\ell]} \theta_l = M\right\}$$

$$= \mathbb{P}\left\{\mathcal{E}_1 = \cdots = \mathcal{E}_{\ell-1} = 1 | \theta_\ell \geq \rho; \ \sum_{l \in [\ell]} \theta_l = M\right\}$$

$$= \mathbb{P}\left\{\mathcal{E}_1 = \cdots = \mathcal{E}_{\ell-1} = 1 | \sum_{l \in [\ell-1]} \theta_l \leq M - \rho; \ \sum_{l \in [\ell]} \theta_l = M\right\} \quad (17)$$

Now it is clear that the probability given in (16) is upper bounded by the one given in (17), since in both cases we assume the same amount of hits in the links of $[\ell]$, but for (16) we assume that more hits (than those assumed by (17)) regard links of $[\ell-1]$. This implies (15) because it holds for any total number of hits in the links of $[\ell]$, and thus for any vector of coalitional hits in $[\ell]$.

Using (15), we obtain that for every $\ell \in \{2, \ldots, m\}$,

$$\mathbb{P}\{\mathcal{E}_1 = \cdots = \mathcal{E}_{\ell-1} = 1\} = \mathbb{P}\{\mathcal{E}_1 = \cdots = \mathcal{E}_{\ell-1} = 1 | \mathcal{E}_\ell = 1\} \cdot \mathbb{P}\{\mathcal{E}_\ell = 1\}$$

$$+ \mathbb{P}\{\mathcal{E}_1 = \cdots = \mathcal{E}_{\ell-1} = 1 | \mathcal{E}_\ell = 0\} \cdot \mathbb{P}\{\mathcal{E}_\ell = 0\}$$

$$\geq \mathbb{P}\{\mathcal{E}_1 = \cdots = \mathcal{E}_{\ell-1} = 1 | \mathcal{E}_\ell = 1\} \quad (18)$$

Therefore,

$$\mathbb{P}\{\mathcal{E}_1 = \cdots = \mathcal{E}_m = 1\} = \mathbb{P}\{\mathcal{E}_1 = \cdots = \mathcal{E}_{m-1} = 1 | \mathcal{E}_m = 1\} \cdot \mathbb{P}\{\mathcal{E}_m = 1\}$$

$$\leq \mathbb{P}\{\mathcal{E}_1 = \cdots = \mathcal{E}_{m-1} = 1\} \cdot \mathbb{P}\{\mathcal{E}_m = 1\} \leq \cdots \leq \prod_{\ell=1}^{m} \mathbb{P}\{\mathcal{E}_\ell = 1\}$$

where the inequalities are due to a recursive application of (18). □

Lemma 5 establishes the negative association of the link lightness events and concludes the proof of the lower bound. □

The Upper Bound. In this last subsection we establish an asymptotically tight upper bound on the Price of Anarchy for the coalitional game on identical parallel links. The approach is similar to the approach of the previous section that bounds the price of anarchy in weighted congestion games on networks with linear delays. More technical details can by found in [14].

Theorem 6. *For every NE* $\mathbf{p}$, $\mathsf{SC}(\mathbf{p}) \leq O(\min\{k, \frac{\log m}{\log \log m}\})\mathsf{OPT}$.

References

1. Althöfer, I.: On sparse approximations to randomized strategies and convex combinations. Linear Algebra and Applications 199, 339–355 (1994)
2. Awerbuch, B., Azar, Y., Epstein, A.: The price of routing unsplittable flow. In: Proc. of the 37th ACM Symp. on Th. of Comp (STOC '05), pp. 57–66. ACM Press, New York (2005)
3. Bazaraa, M.S., Sherali, H.D., Shetty, C.M.: Nonlinear Programming: Theory and Algorithms, 2nd edn. John Wiley and Sons, Inc. Chichester (1993)

4. Beckmann, M., McGuire, C.B., Winsten, C.B.: Studies in the Economics of Transportation. Yale University Press (1956)
5. Chen, X., Deng, X.: Settling the complexity of 2-player nash equilibrium. In: Proc. of the 47th IEEE Symp. on Found. of Comp. Sci (FOCS '06), pp. 261–272. IEEE Computer Society Press, Los Alamitos (2006)
6. Chen, X., Deng, X., Teng, S.H.: Computing nash equilibria: Approximation and smoothed complexity. In: Proc. of the 47th IEEE Symp. on Found. of Comp. Sci (FOCS '06), pp. 603–612. IEEE Computer Society Press, Los Alamitos (2006)
7. Christodoulou, G., Koutsoupias, E.: The Price of Anarchy of Finite Congestion Games. In: Proc. of the 37th ACM Symp. on Th. of Comp. (STOC '05), pp. 67–73. ACM Press, New York (2005)
8. Daskalakis, C., Mehta, A., Papadimitriou, C.: A note on approximate equilibria. In: Spirakis, P.G., Mavronicolas, M., Kontogiannis, S.C. (eds.) WINE 2006. LNCS, vol. 4286, pp. 297–306. Springer, Heidelberg (2006)
9. Dorn, W.S.: Duality in quadratic programming. Quarterly of Applied Mathematics 18(2), 155–162 (1960)
10. Dubhashi, D., Ranjan, D.: Balls and bins: A study in negative dependence. Random Structures & Algorithms 13(2), 99–124 (1998)
11. Fabrikant, A., Papadimitriou, C., Talwar, K.: The complexity of pure nash equilibria. In: Proc. of the 36th ACM Symp. on Th. of Comp (STOC '04), ACM Press, New York (2004)
12. Fotakis, D., Kontogiannis, S., Spirakis, P.: Selfish unsplittable flows. In: Díaz, J., Karhumäki, J., Lepistö, A., Sannella, D. (eds.) ICALP 2004. LNCS, vol. 3142, pp. 226–239. Springer, Heidelberg (2004)
13. Fotakis, D., Kontogiannis, S., Spirakis, P.: Symmetry in network congestion games: Pure equilibria and anarchy cost. In: Erlebach, T., Persinao, G. (eds.) WAOA 2005. LNCS, vol. 3879, pp. 161–175. Springer, Heidelberg (2006)
14. Fotakis, D., Kontogiannis, S., Spirakis, P.: Atomic congestion games among coalitions. In: Bugliesi, M., Preneel, B., Sassone, V., Wegener, I. (eds.) ICALP 2006. LNCS, vol. 4051, pp. 572–583. Springer, Heidelberg (2006)
15. Hoeffding, W.: Probability inequalities for sums of bounded random variables. Journal of the American Statistical Association 58(301), 13–30 (1963)
16. Kamath, A., Motwani, R., Palem, K., Spirakis, P.: Tail bounds for occupancy and the satisfiability threshold conjecture. Random Structures & Algorithms 7(1), 59–80 (1995)
17. Kontogiannis, S., Panagopoulou, P., Spirakis, P.: Polynomial algorithms for approximating nash equilibria in bimatrix games. In: Spirakis, P.G., Mavronicolas, M., Kontogiannis, S.C. (eds.) WINE 2006. LNCS, vol. 4286, pp. 286–296. Springer, Heidelberg (2006)
18. Kontogiannis, S., Spirakis, P.: Well supported approximate equilibria in bimatrix games: A graph theoretic approach. In: Proc. of the 32nd Int. Symp. on Math. Found. of Comp. Sci (MFCS '07). Preliminary version appeared as technical report DELIS-TR-0487 of EU Project DELIS – Dynamically Evolving Large Scale Information Systems, January 2007. LNCS, Springer, Heidelberg (2007)
19. Koutsoupias, E., Mavronicolas, M., Spirakis, P.: Approximate equilibria and ball fusion. Theory of Computing Systems 36(6), 683–693 (2003) (Special Issue devoted to SIROCCO'02)
20. Koutsoupias, E., Papadimitriou, C.: Worst-case equilibria. In: Meinel, C., Tison, S. (eds.) STACS 99. LNCS, vol. 1563, pp. 404–413. Springer, Heidelberg (1999)

21. Lipton, R., Markakis, E., Mehta, A.: Playing large games using simple strategies. In: Proc. of the 4th ACM Conf. on El. Comm (EC '03). Assoc. of Comp. Mach, pp. 36–41. ACM Press, New York (2003)
22. Monderer, D., Shapley, L.: Potential games. Games & Econ. Behavior 14, 124–143 (1996)
23. Nash, J.: Noncooperative games. Annals of Mathematics 54, 289–295 (1951)
24. Papadimitriou, C., Steiglitz, K.: Combinatorial Optimization: Algorithms and Complexity. Prentice-Hall, Inc. Englewood Cliffs (1982)
25. Rosenthal, R.W.: A class of games possessing pure-strategy nash equilibria. Int. J. of Game Theory 2, 65–67 (1973)
26. Roughdarden, T., Tardos, E.: How bad is selfish routing? J. of ACM 49(2), 236–259 (2002)

Randomized Algorithms and Probabilistic Analysis in Wireless Networking*

Aravind Srinivasan

Department of Computer Science and Institute for Advanced Computer Studies,
University of Maryland, College Park, MD 20742, USA

Abstract. Devices connected wirelessly, in various forms including computers, hand-held devices, *ad hoc* networks, and embedded systems, are expected to become ubiquitous all around us. Wireless networks pose interesting new challenges, some of which do not arise in standard (wired) networks. This survey discusses some key probabilistic notions – both randomized algorithms and probabilistic analysis – in wireless networking.

1 Introduction

It is anticipated that wireless networking will continue to have considerable growth in the foreseeable future, and that devices connected in wireless fashion will pervade our world. As compared to wired networks, two particular challenges arise in the wireless setting: *energy-conservation* (since tiny and embedded devices, especially, have very little access to a continuous source of power) and *interference* between nearby transmissions. This survey will briefly reference certain algorithmic approaches and modeling/analysis techniques that have been developed to tackle such issues. This is certainly not meant to be an encyclopedic survey, and many key papers will not be referred to here. Rather, we hope to spur the interest of the reader in exploring this exciting area further.

It is natural that probabilistic considerations should help in our present context. First, in addition to their well-known advantages (such as leading to faster and simpler algorithms), randomized algorithms play a powerful role in any type of *distributed* system, through paradigms such as symmetry-breaking; a natural example of this is contention resolution at the MAC (Media Access Control) layer of wireless networks, where nearby radios try to access their local radio spectrum in a contention-free manner for short periods of time. Second, probabilistic analysis is useful as always in determining "typical" properties of systems. Consider the example of a set of n transceivers distributed randomly in a bounded region: this is a particularly natural way to model a set of n sensors thrown over the region. Probabilistic models are also one obvious approach to model mobility.

We will reference below a few representative examples of randomized algorithms and probabilistic analysis in wireless networking. The reader is referred to [22] for a comprehensive tutorial on algorithms for sensor networks.

* This research has been supported in part by NSF ITR Award CNS-0426683 and NSF Award CNS-0626636.

J. Hromkovič et al. (Eds.): SAGA 2007, LNCS 4665, pp. 54–57, 2007.
© Springer-Verlag Berlin Heidelberg 2007

2   Four Topics

We briefly consider four topics: (a) energy conservation, (b) design and analysis of media-access protocols, (c) probabilistic analysis of network capacity assuming a random network topology and/or a random traffic matrix, and (d) probabilistic analysis/randomized algorithms in efficient aggregation of information in sensor networks.

A candidate approach that has been proposed for energy savings is through co-operation: subsets of nodes act as a *backbone* while other nodes go to sleep, and the backbone stores and forwards messages for all nodes during this period [5]. This backbone is updated frequently, in the interest of fairness. Graph-theoretically, such problems are closely related to connected domination (which is a natural condition to impose on the backbone), domatic partitions (the problem of partitioning the network into backbones) etc. Randomization plays an essential role in much work in this area; see, e.g., [5,8,6]. However, good *deterministic* deterministic approximation algorithms are also possible, in the case where the underlying network satisfies inter-node distances that approximate those in some low-dimensional Euclidean space (e.g., if the network has small *doubling-dimension*) [19]. However, what if the nodes are selfish and will deviate from the protocol if it satisfies their individual selfish desire to conserve their own power? See [16] for a game-theoretic approach to this problem.

Second, as mentioned above, random access is a natural approach for accessing the medium (i.e., the radio spectrum). While this is a classical, well-studied issue (see, e.g., [1,3,11]), the analysis becomes much harder for modern protocols in wireless networking; this is due to additional constraints and features such as routing in the network, the possibility of multi-channel multi-radio transceivers, etc. The works [21,12] respectively analyze protocols such as relatives of IEEE 802.11 and present new random-access protocols for multihop wireless networks. Combined MAC scheduling and end-to-end routing for a given set of end-to-end connections is achieved in [14]; the ideas include linear-programming relaxations, distributed graph-coloring, and geometric arguments. See [4] for a provably-good deterministic distributed packet-scheduling algorithm for the multi-channel multi-radio setting.

Our next two topics largely involve probabilistic analysis.

Our third topic relates to the fact that as opposed to wired networks, the capacity or throughput (measured using maximum flow, maximum concurrent flow, or other similar objective functions) of wireless networks is complicated by the presence of interference. In a seminal paper, Gupta & Kumar considered the capacity of a multi-hop wireless network formed by distributed n transceivers randomly in the unit square; the traffic matrix is random (such as a random permutation of the transceivers), and the interference model can be any one of a few standard candidates [10]. This spurred quite some work on generalizations (e.g., to hybrid networks which have a few base stations, see [17,13]). The capacity has also been approximated in the worst case using geometric arguments [15]. Random-graph models for wireless (sensor) networks have been analyzed in works including [20,9,18].

Finally, we very briefly mention the field of collecting/aggregating information from the sensors in a sensor network, in an energy-efficient manner. The basic idea is that since the data at nearby nodes are likely to be *correlated*, we could try and achieve some sort of information-theoretic compression while transmitting the data from the sensors, as compared to separately outputting the data at each sensor. See [2,7] for two papers in this growing area; interesting further connections to fields such as information theory and machine learning appear ripe for investigation here.

Acknowledgment. I thank V. S. Anil Kumar, David Levin, Madhav Marathe, and Srinivasan Parthasarathy for their help in preparing this article.

References

1. Abramson, N.: The ALOHA system. In: Abramson, N., Kuo, F. (eds.) Computer-Communication Networks, Englewood Cliffs, New Jersey, Prentice Hall, Englewood Cliffs (1973)
2. Adler, M.: Collecting Correlated Information from a Sensor Network. In: Proc. ACM-SIAM Symposium on Discrete Algorithms, pp. 479–488. ACM Press, New York (2005)
3. Aldous, D.: Ultimate instability of exponential back-off protocol for acknowledgement-based transmission control of random access communication channels. In: IEEE Trans. on Information Theory, vol. IT-13, pp. 219–223. IEEE Computer Society Press, Los Alamitos (1987)
4. Brzezinski, A., Zussman, G., Modiano, E.: Enabling distributed throughput maximization in wireless mesh networks: a partitioning approach. Proc. International Conference on Mobile Computing and Networking, pp. 26–37 (2006)
5. Chen, B., Jamieson, K., Balakrishnan, H., Morris, R.: Span: An Energy-Efficient Coordination Algorithm for Topology Maintenance in Ad Hoc Wireless Networks. ACM Wireless Networks 8, 481–494 (2002)
6. Dubhashi, D., Mei, A., Panconesi, A., Radhakrishnan, R., Srinivasan, A.: Fast Distributed Algorithms for (Weakly) Connected Dominating Sets and Linear-Size Skeletons. Journal of Computer and System Sciences 71, 467–479 (2005)
7. Enachescu, M., Goel, A., Govindan, R., Motwani, R.: Scale Free Aggregation in Sensor Networks. In: Proc. First International Workshop on Algorithmic Aspects of Wireless Sensor Networks (Algosensors) (2004)
8. Feige, U., Halldórsson, M.M., Kortsarz, G., Srinivasan, A.: Approximating the Domatic Number. SIAM Journal on Computing 32, 172–195 (2002)
9. Goel, A., Rai, S., Krishnamachari, B.: Sharp thresholds for monotone properties in random geometric graphs. Annals of Applied Probability, 2004 (to appear) (Preliminary version in Proc. ACM Symposium on Theory of Computing)
10. Gupta, P., Kumar, P.R.: The capacity of wireless networks. IEEE Transactions on Information Theory 46, 388–404 (2000)
11. IEEE Trans. on Information Theory. IT-31, special issue (1985)
12. Joo, C., Shroff, N.B.: Performance of Random Access Scheduling Schemes in Multi-Hop Wireless Networks. In: Proc. IEEE International Conference on Computer Communications, pp. 19–27 (2007)
13. Kozat, U.C., Tassiulas, L.: Throughput Scalability of Wireless Hybrid Networks over a Random Geometric Graph. Wireless Networks 11, 435–449 (2005)

14. Kumar, V.S.A., Marathe, M.V., Parthasarathy, S., Srinivasan, A.: End-to-End Packet-Scheduling in Wireless Ad-Hoc Networks. In: Proc. ACM-SIAM Symposium on Discrete Algorithms, pp. 1014–1023 (2004)
15. Kumar, V.S.A., Marathe, M.V., Parthasarathy, S., Srinivasan, A.: Algorithmic Aspects of Capacity in Wireless Networks. In: Proc. ACM International Conference on Measurement and Modeling of Computer Systems, pp. 133–144 (2005)
16. Lee, S., Levin, D., Gopalakrishnan, V., Bhattacharjee, B.: Backbone construction in selfish wireless networks. In: Proc. ACM International Conference on Measurement and Modeling of Computer Systems (2007)
17. Liu, B., Liu, Z., Towsley, D.F.: On the Capacity of Hybrid Wireless Networks. Proc. IEEE INFOCOM (2003)
18. Muthukrishnan, S., Pandurangan, G.: The bin covering technique for thresholding random geometric graph properties. Proc. ACM-SIAM Symposium on Discrete Algorithms, pp. 989–998 (2005)
19. Pemmaraju, S.V., Pirwani, I.A.: Energy conservation via domatic partitions. In: Proc. ACM International Symposium on Mobile Ad Hoc Networking and Computing, pp. 143–154 (2006)
20. Servetto, S.D., Barrenechea, G.: Constrained Random Walks on Random Graphs: Routing Algorithms for Large Scale Wireless Sensor Networks. In: Proc. ACM International Workshop on Wireless Sensor Networks and Applications, held in conjunction with ACM MOBICOM, ACM Press, New York (2002)
21. Sharma, G., Ganesh, A., Key, P.: Performance Analysis of Contention Based Medium Access Control Protocols. In: Proc. IEEE International Conference on Computer Communications, pp. 1–12 (2006)
22. Wattenhofer, H.: Algorithms for Wireless Sensor Networks (Tutorial). In: Proc. European Workshop on Wireless Sensor Networks (2006)

A First Step Towards Analyzing the Convergence Time in Player-Specific Singleton Congestion Games[*]

Heiner Ackermann

Department of Computer Science
RWTH Aachen, D-52056 Aachen, Germany
ackermann@cs.rwth-aachen.de

Abstract. We initiate studying the convergence time to Nash equilibria in player-specific singleton congestion games. We consider simple games that have natural representations as graphs as we assume that each player chooses between two resources. We are not able to present an analysis for general graphs. However, we present first results for interesting classes of graphs. For the class of games that are represented as trees, we show that every best-response schedule terminates after $O(n^2)$ steps. We also consider games that are represented as circles. We show that deterministic best response schedules may cycle, whereas the random best response schedule, which selects the next player to play a best response uniformly at random, terminates after $O(n^2)$ steps in expectation. These results imply that in player-specific congestion games in which each player chooses between two resources, and each resource is allocated by at most two players, the random best response schedule terminates quickly. Our analysis reveals interesting relationships between random walks on lines and the random best response schedule.

1 Introduction

In this paper, we take a first step towards analyzing the convergence time in player-specific singleton congestion games. In such games, we are given a set of resources and a set of players. Each player is equipped with a set of non-decreasing, player-specific delay functions which measure the delay the player would experience from allocating a particular resource while sharing it with a certain number of other players. A player's goal is to allocate a *single* resource with minimum delay given fixed choices of the other players. As every such player-specific singleton congestion game possesses a Nash equilibrium [9], we are interested in analyzing the maximum number of steps until players iteratively changing to resources with minimum delay reach a Nash equilibrium. In the following, we call this process the *best response dynamics*. Furthermore, we call a schedule which selects the next player to play a best response a *best response schedule*. If all players have identical delay functions, that is, if all players sharing a resource observe the same delay, we omit the term player-specific and call such a game a standard singleton congestion game. In the case of standard singleton congestion games, Ieong et al. [7] show that the best response dynamics terminates after at most n^2m steps

---

[*] This work was supported in part by the EU within the 6th Framework Programme under contract 001907 (DELIS).

J. Hromkovič et al. (Eds.): SAGA 2007, LNCS 4665, pp. 58–69, 2007.
© Springer-Verlag Berlin Heidelberg 2007

in a Nash equilibrium. Here, n equals the number of players, and m the number of resources. Their analysis relies on a potential functions that strictly decreases whenever a player plays a best response.

Milchtaich [9] observes that player-specific singleton congestion games do not admit a potential function as there exist games in which the best response dynamics may cycle. However, he proves that from every state of such a game, there exists a sequence of best responses of polynomial length leading to a Nash equilibrium. He concludes that if we select the next player to play a best response uniformly at random, the *random best response schedule* terminates in a Nash equilibrium after a finite number of steps with probability one. His analysis leaves open the question how long it takes until the random best response schedule terminates. In this paper, we address this question as we think that it is an important and interesting one.

Currently, we are not able to analyze the convergence time in arbitrary player-specific singleton congestion games. However, we begin with very simple yet interesting classes of games, and consider games in which each player chooses between two alternatives. These games can be represented as graphs: each resource corresponds to a node, each player to an edge. In the following, we call games that can be represented as graphs with topology t player-specific congestion games on topology t. We consider games on trees and circles. In the case of player-specific congestion games on trees we show that the best response dynamics cannot cycle. In order to prove this we observe that one can replace the player-specific delay functions by common delay functions without changing the players preferences. Thus, player-specific congestion games on trees are isomorphic to standard congestion games on tress. From this observation we conclude a tight upper bound of $\Theta(n^2)$ on the convergence time in such games. We proceed with player-specific congestion games on circles, and show that these games are in some sense the simplest games in which the best response dynamics may cycle. As we are only given four different delay values per player, we characterize with respect to the ordering of these four values in which cases the best response dynamics may cycle and when not. We observe that the delay functions have to be chosen in the right way in order to obtain games in which deterministic best response schedules may cycle. Finally, we analyze the convergence time of the random best response schedule in such games, and prove a tight bound of $\Theta(n^2)$. Our analysis reveals interesting relationships between random walks on lines and the random best response schedule.

1.1 Definitions and Notations

Player-specific Singleton Congestion Games: A *player-specific singleton congestion game* Γ is a tuple $(\mathcal{N}, \mathcal{R}, (\Sigma_i)_{i \in \mathcal{N}}, (d_r^i)_{r \in \mathcal{R}}^{i \in \mathcal{N}})$ where $\mathcal{N} = \{1, \ldots, n\}$ denotes the set of players, $\mathcal{R} = \{1, \ldots, m\}$ the set of resources, $\Sigma_i \subseteq \mathcal{R}$ the strategy space of player i, and $d_r^i \colon \mathbb{N} \to \mathbb{N}$ a strictly increasing player-specific delay function associated with player i and resource r. In the following, we assume that for every player i, every pair of resources $r_1, r_2 \in \Sigma_i R$, and every pair $n_{r_1}, n_{r_2} \in \mathbb{N}$: $d_{r_1}^i(n_{r_1}) \neq d_{r_2}^i(n_{r_2})$. The reason for this assumption will be explained later.

We denote by $S = (r_1, \ldots, r_n)$ the *state of the game* where player i allocates resource $r_i \in \Sigma_i$. For a state S, we define the *congestion* $n_r(S)$ on resource r by $n_r(S) = |\{i \mid r = r_i\}|$, that is, $n_r(S)$ equals the number of players sharing resource r in state S. We assume that players act selfishly seeking to allocate single resources minimizing their individual delays. The delay of player i from allocating resource r in state S is given by $d_r^i(n_r(S))$. Given a state $S = (r_1, \ldots, r_n)$, we call a resource $r^* \in \Sigma_i \setminus \{r_i\}$ a *best response* of player i to S if, for all $r' \in \Sigma_i \setminus \{r_i\}$, $d_{r^*}^i(n_{r^*}(S) + 1) \leq d_{r'}^i(n_{r'}(S) + 1)$, and if $d_{r^*}^i(n_{r^*}(S) + 1) < d_{r_i}^i(n_{r_i}(S))$. Note that due to our assumptions on the delay functions, best responses are unique. The standard solution concept to player-specific singleton congestion games are *Nash equilibria*. A state S is a Nash equilibrium if no player has an incentive to allocate another resource.

In this paper, we only consider games that have natural representations as graphs. We assume that each player chooses between two resources and that no two players choose between the same two resources. In this case, we can represent the resources of such a game as the nodes of a graph and the players as the edges. The direction of an edge naturally corresponds to the strategy the player plays. We call games that can be represented as graphs with topology t player-specific singleton congestion games on topology t.

In the following, we will sometime refer to *standard singleton congestion games*. Standard singleton congestion games are defined in the same way as player-specific singleton congestion games except that we are not given player-specific delay functions $d_r^i, r \in \mathcal{R}, i \in \mathcal{N}$, but common delay functions $d_r, r \in \mathcal{R}$.

Transition Graph: We define the *transition graph* $TG(\Gamma)$ of a player-specific singleton congestion game Γ as the graph that contains a vertex for every state of the game. Moreover, there is a directed edge from state S to state S' if we would obtain S' from S by permitting one player to play a best response.

Best Response Dynamics and Best Response Schedule: We call the dynamics in which players iteratively play best responses given fixed choices of the other players *best response dynamics*. Furthermore, we use the term *best response schedule* to denote an algorithm that selects given a state S the next player to play a best response. We assume that such a player is always selected among those players who have an incentive to change their strategy. The convergence time $t(n, m)$ of a best response schedule is the maximum number of steps to reach a Nash equilibrium in any game with n players and m resources independent of the initial state. If the schedule is a randomized algorithm then $t(n, m)$ refers to the expected convergence time.

The Type of a Player: Ieong et al. [7] consider standard singleton congestion games. They observe that one can always replace the delay values $d_r(n_r)$ with $r \in \mathcal{R}$ and $1 \leq n_r \leq n$ by their ranks in the sorted list of these values without changing the best response dynamics. Note that this approach is not restricted to standard singleton congestion games but also applies to player-specific singleton congestion games. That is, given a player-specific congestion game Γ, fix a player i and consider a list of all delays $d_r^i(n_r)$ with $r \in \mathcal{R}$ and $1 \leq n_r \leq n$. Assume that this list is sorted in a non-decreasing order. For each resource r, we define an alternative player-specific delay function $\tilde{d}_r^i : \mathbb{N} \rightarrow \mathbb{N}$ where, for each possible congestion n_r, $\tilde{d}_r^i(n_r)$ equals the rank of the delay $d_r^i(n_r)$ in the aforementioned list of all delays. Due to our assumptions on the delay functions, all ranks are unique. In the following, we define the *type of a player i* by the ordering of the player-specific delays $d_r^i(1), \ldots, d_r^i(n)$ of the resources $r \in \Sigma_i$.

1.2   Related Work

Milchtaich [9] introduces player-specific singleton congestion games, and proves that every such game possesses a Nash equilibrium if the player-specific delay functions are non-decreasing. He also observes that player-specific singleton congestion game do not admit a potential function as there exist games in which the best response dynamics may cycle. Milchtaich also observes that from every state of such a game there exists a sequence of best responses leading to a Nash equilibrium. As these sequences can be computed efficiently, there exists a polynomial time algorithm computing Nash equilibria in player-specific singleton congestion games. Ackermann, Röglin, and Vöcking [2] extend these results to player-specific matroid congestion games. In such games the players' strategy spaces are sets of bases of matroids on the resources.

A model closely related to player-specific congestion games are standard congestion games. Rosenthal [10] introduces these games and proves with a potential function that every such game, regardless of the players' strategy spaces, and of any assumptions on the delay functions, possesses a Nash equilibrium. Ieong et al. [7] address the convergence time in such games. They consider standard singleton congestion games, and show that the best response dynamics converges quickly. Fabrikant, Papadimitriou, and Talwar [5] show that in general standard congestion games players do not convergence quickly. Their result especially holds in the case of network congestion games, in which players seek to allocate a path between different source-sink pairs. Later, Ackermann, Röglin, and Vöcking [1] extended the result of Ieong et al. [7] towards matroid congestion games, and prove that the matroid property is the maximal property on the players' strategy spaces guaranteeing polynomial time convergence.

There are several other articles addressing the convergence time to Nash equilibria in standard or weighted singleton congestion games [3,4,6]. All these analyses depend on potential functions. To our knowledge, this is the first paper addressing the convergence time in player-specific singleton congestion games which do not possess a potential function.

2   Games on Trees

In this section, we consider player-specific congestion games on trees. First, we observe that one can always replace the player-specific delay functions by common delay functions such that the players' types are preserved. In this case, we obtain a standard singleton congestion game, whose transition graph equals the transition graph of the player-specific game. We conclude the following theorem.

Theorem 1. *Let Γ be a player-specific congestion game on a tree. Then the transition graph of Γ is cycle-free.*

A formal proof of Theorem 1, can be found in a full version. Furthermore, by reworking the proof of the convergence time of standard congestion game [7], we conclude that every best response schedule for player-specific congestion games on trees terminates after $O(n^2)$ steps. We also like to mention that this upper bound is tight. That is, there exists an infinite family of instances of player-specific congestion games on trees and a best response schedule that terminates after $\Omega(n^2)$ steps on every instance of the family if the initial state is chosen appropriately. A precise description of these instances and of the corresponding schedule can be found in the full version, too.

Corollary 2. *Let Γ be a player-specific congestion game on a tree. Then every best response schedule terminates after $O(n^2)$ steps. Moreover, this analysis is tight.*

3   Player-Specific Congestion Games on Circles

We now consider player-specific congestion games Γ on circles. Without loss of generality, we assume that for every player i: $\Sigma_i = \{r_i, r_{i+1 \mod n}\}$. In the following, we call r_i the 0- and $r_{i+1 \mod n}$ the 1-strategy of player i. Furthermore, we drop the mod n terms and assume that all indices are computed modulo n. Due to our assumptions on the delay functions, there are six different types of players in such games.

$$d_{r_i}^i(1) < d_{r_i}^i(2) \quad < d_{r_{i+1}}^i(1) < d_{r_{i+1}}^i(2) \qquad \text{type 1}$$
$$d_{r_i}^i(1) < d_{r_{i+1}}^i(1) < d_{r_i}^i(2) \quad < d_{r_{i+1}}^i(2) \qquad \text{type 2}$$
$$d_{r_i}^i(1) < d_{r_{i+1}}^i(1) < d_{r_{i+1}}^i(2) < d_{r_i}^i(2) \qquad \text{type 3}$$

We call the other 3 types, which can be obtained by exchanging the identities of the resources r_i and r_{i+1} in the above inequalities, type 1', type 2', and type 3'. Furthermore, we call two players i, j *consecutive*, if they share a resource, that is, if $j = i + 1$ or $i = j + 1$. Given a state S, we call two consecutive players *synchronized*, if both play the same strategy, that is, if both either play their 0- or their 1-strategy. Moreover, we call a set of consecutive players $i, \ldots, j$ synchronized if all players play the same strategy.

3.1   A Lower Bound

As a first step, we present an infinite family of games possessing cycles in their transition graphs, and show a lower bound of $\Omega(n^2)$ on the convergence time of the random best response schedules on these games.

 Consider the family of games on a circle such that all players are of type 3. It is not difficult to verify that in every Nash equilibrium of such a game all players are synchronized. Let S be a state with the following properties. In S there are two non-empty sets S_0 and S_1 of synchronized players. Players in S_0 all play their 0-strategy, whereas players in S_1 all play their 1-strategy. Again, it is not difficult to verify that in every such state there are two players who have an incentive to change their strategies. From both sets only the first player clockwise has an incentive to change her strategy. From this observation we conclude that there exist cycles in the transition graphs of such games. We obtain such a cycle by selecting players from the two sets alternately, and letting them play best responses.

 In order to prove a lower bound on the random best response schedule, observe that with probability $1/2$ the total number of players playing their 0-strategy increases or decreases by one whenever a player is selected uniformly at random. After the strategy change either all players are synchronized, and therefore the random best response schedule terminates, or again we are in a state S' with two sets of synchronized players. Observe now that this process is isomorphic to a random walk on a line of nodes $v_0, \ldots, v_n$. The node v_i corresponds to the fact that i players play their 0-strategy. As the expected time of a random walk on a line with $n + 1$ nodes to reach one of the two ends of the line is $\Theta(n^2)$ if the walk starts in the middle of the line [8], we obtain a lower bound of $\Omega(n^2)$ on the convergence time of player-specific congestion games on circles.

Corollary 3. *There exists a family of instances of player-specific congestion games on circles with initial states such that the convergence time of the random best response schedule is lower bounded by $\Omega(n^2)$.*

In the following sections, we present a matching upper bound on the convergence time of the random best response schedule, and prove the following theorem which follows from Lemma 5, 8, 9, 10 and 11.

Theorem 4. *Let Γ be a player-specific congestion game on a circle. Then the random best response schedule terminates after $O(n^2)$ steps in expectation. Moreover, this analysis is tight.*

We characterize with respect to the types of the players in which cases there are cycles in the transition graphs of such games. We show that in almost all cases there are no cycles; cycles only exist if all players are of type 3 or type 3'. We analyze the convergence time of deterministic best response schedules in cycle-free games by developing a general framework that allows to derive potential functions. Finally, we analyze the convergence time of the random best response schedule in the case of games with players of type 3 or type 3'.

3.2 The Impact of Type 1 Players

In this section, we investigate the impact of type 1 players on the existence of cycles in the transition graphs and on the convergence time of best response schedules. An intuitive argument for the absence of cycles in the transition graphs of games with at least one player of type 1 is that every player of type 1 changes her strategy at most once, whereas in a cycle every player changes her strategy at least two times.

Lemma 5. *Let Γ be a player-specific congestion game on a circle. If there exists at least one player of type 1, then $TG(\Gamma)$ is cycle-free. Moreover, every best response schedule terminates after $O(n^2)$ steps.*

The proof can be found in the the full version. The running time follows since one can always split the game into two player-specific congestion games on trees, and embed the transition graphs of these two games into the transition graph of Γ. In the first game the player of type 1 is fixed to her 0-strategy, in the second one to her 1-strategy. Note that Lemma 5 also holds in the case of a player of type 1'. In the following sections, we will therefore assume that there exist no players of type 1 or 1', as otherwise we could apply Lemma 5.

3.3   A Framework to Analyze the Convergence Time

In this section, we present a framework to analyze the convergence time of best response schedules in player-specific congestion games on circles. Let Γ be a game such that there is no player of type 1 or 1'. First, we investigate whether there is a sufficient condition such that player i does not want to change her strategy in a state S of Γ.

Observation 6. *Suppose that player i is not of type 1 or 1'. Then if she is synchronized with the players $i - 1$ and $i + 1$ in S, she has no incentive to change her strategy.*

In the following, we call a resource r *overloaded* in state S, if two players share r. Additionally, we call a resource r' *underloaded* in state S, if no player allocates r'. Obviously in every state of Γ, the total number of overloaded resources equals the total number of underloaded resources. From Observation 6, we conclude that in every state S only players who allocate a resource that is currently overloaded or who could allocate a resource that is currently underloaded might have an incentive to change their strategy.

Based on this observation, we now present a general framework to analyze the convergence time of best response schedules. First, we introduce the notion of *over- and underload tokens*. Given an arbitrary state S of Γ, we place an *overload token* on every overloaded resource. Additionally, we place an *underloaded token* on every underloaded resource. Obviously over- and underload tokens alternate on the circle. Furthermore, note that a legal placement of tokens uniquely determines the strategies the players play. A placement of tokens is legal if no two tokens share a resource, and if the tokens alternate on the circle.

In the following, we investigate in which directions tokens move if players play best responses. Consider first a sequence of resources $r_i, \ldots, r_j$ and assume that players $i, \ldots, j - 1$ are of the same type t. Additionally, assume that an overload token is placed on resource r_k, and that an underload token is placed on resource r_l with $i < k < l < j$. The scenario we consider is depicted in Figure 1.

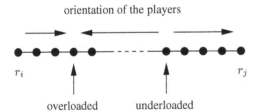

	overload	underload
type 2	anticlockwise	clockwise
type 2'	clockwise	anticlockwise
type 3	clockwise	clockwise
type 3'	anticlockwise	anticlockwise

Fig. 1. In which directions do the tokens move?

Assume first, that the distance (number of edges) between the two tokens is at least two. Thus, $|l - k| \geq 2$. In this case, we observe that each token can only move in one direction. The directions are uniquely determined by the type of the players. They can be derived from investigating, with respect to the players' type t, which players have incentives to change their strategy. The directions are stated in Figure 1, too. Assume now that the distance between the two tokens is one. That is, $k = l - 1$. Thus, there exists a player who is interested in the over- and underloaded resource, and who currently allocates the overloaded one. It is not difficult to verify that this player always has an incentive to change her strategy. Note that this holds regardless of the player's type since we assumed that there are no players of type 1 and 1'. Observe now that after the strategy change of this player all players are synchronized and therefore there exist no over- and underloaded resources anymore. In the following, we call such an event a *collision of tokens*.

So far, we considered sequences of players of the same type and observed that there is a unique direction in which tokens of the same kind move. In sequences with multiple types of players such unique directions do not exist any longer, i.e., overload as well as underload tokens can move in both directions. However, if two players of different types share a resource and if due to best responses of both players an over- or underload token moves onto this resource, then the token could stop there. In the following, we formalize this observation with respect to overload tokens and introduce the notion of *termination points*.

Definition 7. *We call a resource r_i a termination point of an overload token if the following conditions are satisfied.*

1. *The players $i - 1$ and i have different types. Let these types be t_{i-1} and t_i.*
2. *In sets of consecutive players of type t_{i-1} overload tokens move clockwise, whereas they move anticlockwise in sets of consecutive players of type t_i.*

We illustrate the definition in Figure 2a). Let player $i - 1$ be of type 3, and let player i be of type 2. In this case, the requirements of the definition are satisfied. Assume, that the player $i - 1$ plays her 1-strategy and that she is synchronized with the player $i - 2$. Additionally, assume that player i plays her 0-strategy and that she is synchronized with the player $i + 1$. Observe now that the token cannot move as neither the player $i - 1$ nor the player i has an incentive to change her strategy. Suppose now that initially all players along the path play their 0-strategy. Then an overload token that moves from the left to the right along the path stops at r_i. The token may only move on if one of the two players is not synchronized with its neighbor any longer. In this case, this player always has an incentive to change her strategy as she can allocate a resource that is currently underloaded. Thus, an underload and an overload token collide. Additionally, if initially all players play their 1-strategy and an overload token moves from the right to the left along the path, we observe the same phenomenon. The token cannot pass the resource r_i unless it collides with an underload token.

Note that the definition of a termination point can easily be adopted to underload tokens. A list of all termination point is given in Figure 2b). In the left column we present all termination points for overload tokens, in the right one for underload tokens.

3.4 Analyzing the Convergence Time

In this section, we analyze the convergence time in player-specific congestion games on circles. We distinguish between the following four cases.

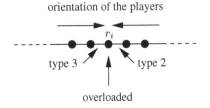

→→←←	←←→→
2' 2	2 2'
3 3'	3 3'
3 2	-
-	2 3'
-	3 2'
2' 3'	-

Fig. 2. a) An example of a termination point.   b) A list of all termination points.

Case 1: For both kinds of tokens there exists at least one termination point.

Case 2: Only for one kind of token there exists at least one termination point.

Case 3: There exist no termination points but over- and underload tokens move in opposite directions.

Case 4: There exist no termination points and over- and underload tokens move in the same direction.

In the first two cases, we present potential functions and prove that the transition graphs of such games are cycle-free and that every best response schedule terminates after $O(n^2)$ steps. In the third case, we can do slightly better and prove an upper bound of $O(n)$ on the convergence time. Note that in all cases one can easily construct matching lower bounds. Only in the fourth case deterministic best response schedules may cycle. In this case, we prove that the random best response schedule terminates after $O(n^2)$ steps in expectation.

Before we take a closer look at the different cases, we discuss which games with respect to their players' types belong to which case. Games with players of type 2 and 2' or with players of type 3 and 3' belong to the first case. Additionally, some games with more than two types of players belong to this case. The second case covers all games with at least three different kind of players which do not belong to the first case. Furthermore, it covers games with type 2 and type 3 players, and games with type 2' and type 3' players. Games with type 2 players only, or games with type 2' players only belong to the third case. Finally, games with type 3 players only and games with type 3' players only belong to the fourth case. These observations can easily be derived from Figure 2 b).

Case 1

Lemma 8. *Let Γ be a player-specific congestion game on a circle such that there are termination points for both kinds of tokens. Then $TG(\Gamma)$ is cycle-free. Moreover, every best response schedule terminates after $O(n^2)$ steps.*

Proof. Let S be a state of Γ and consider the mapping that maps every token in S to the next termination point lying in the direction in which the token moves. In the following, we define $d(t, S)$ as the distance of a token t in state S to its corresponding termination point. Obviously $d(t, S) \leq n$. Consider now the potential function $\phi(S) = \sum_{\text{token } t} d(t, S)$ and suppose that a player plays a best response. Then either one tokens moves closer to its termination point or two tokens collide. In both cases $\phi(S)$ decreases by at least 1. Thus, $\phi(S)$ strictly decreases if a player plays a best response and therefore, $TG(\Gamma)$ is cycle-free. Moreover, as $\phi(S)$ is upper bounded by $O(n^2)$, every best response schedule terminates after $O(n^2)$ steps.   □

Case 2

Lemma 9. *Let Γ be a player-specific congestion game on a circle such that there are termination points only for one kind of token. Then $TG(\Gamma)$ is cycle-free. Moreover, every best response schedule terminates after $O(n^2)$ steps.*

Proof. Without loss of generality, assume that termination points only exist for overload tokens. In this case, we define $d(t_o, S)$ for every overload token t_o as in the proof of Lemma 8. For every underload token t_u we define $d(t_u, S)$ as follows. Let t_o be the first overload token lying in the same direction as t_u moves.

1. If t_o moves in the opposite direction as t_u, then we define $d(t_u, S)$ as the distance between the two tokens. The distance of two tokens moving in opposite directions is defined as the number of moves of these tokens until they collide.
2. If t_o moves in the same directions as t_u, then we define $d(t_u, S)$ as the distance between t_u and t_o plus the distance between t_o and the first termination point at which t_o has to stop. Thus, $d(t_u, S)$ equals the maximum number of moves of these two tokens until they collide.

Observe, that for every underload token t_u: $d(t_u, S) \leq 2n$. Consider, the potential function $\phi \colon \Sigma \to \mathbb{N} \times \mathbb{N}$ with $\phi(S) = (\phi_1(S), \phi_2(S))$. $\phi_1(S)$ equals the total number of overload tokens in S, whereas $\phi_2(S)$ equals the sum of all $d(t, S)$ for all under- and overload tokens. Suppose now that a player plays a best response. Obviously if two tokens collide, then $\phi_1(S)$ decrease by one. Moreover, if there is no collision, then $\phi_2(S)$ decreases. Note that in the first case ϕ_2 may increase. This may happen if, due to the collision, $d(t_u, S)$ of a remaining underload token t_u has to be recomputed as its associated overload token has been removed. The new value is upper bounded by the sum of the old values of t_u and the collided underload token plus 1. Now consider an ordering $<_\phi$ of the states of Γ with respect to ϕ. Let S and S' be two states of Γ. Then

$$S <_\phi S' \Leftrightarrow \begin{cases} \phi_1(S) < \phi_1(S') & \text{or} \\ \phi_1(S) = \phi_1(S') \text{ and } \phi_2(S) < \phi_2(S') . \end{cases}$$

Observe, that ϕ strictly decreases if a player plays a best response. Thus, $TG(\Gamma)$ is cycle-free. Additionally, observe that ϕ_1 is upper bounded by n, and that ϕ_2 is upper bounded by n^2. However, as ϕ_2 only increases by one when ϕ_1 decreases, we conclude that every best response schedule terminates after $O(n^2)$ steps.   □

Case 3. The analysis of the third case follows similar arguments then the proofs of Lemma 8 and Lemma 9. We therefore omitted a formal proof of the following lemma.

Lemma 10. *Let Γ be a player-specific congestion game on a circle such that there exist no termination points but over- and underload tokens move in opposite directions. Then $TG(\Gamma)$ is cycle-free. Moreover, every best response schedule terminates after $O(n)$ steps.*

Case 4. In the following, we present a proof of the fourth case with respect to players of type 3. By symmetry of the types 3 and 3', the same result holds with respect to games with players of type 3', too.

Lemma 11. *Let Γ be a player-specific congestion game on a circle such that all players are of type 3. Then the random best response schedule terminates after $O(n^2)$ steps in expectation.*

Proof. In order to prove the lemma, we prove the following lemma.

Lemma 12. *In every state S of Γ the number of players who want to change from their 0- to their 1-strategy equals the number of players who want to change from their 1- to their 0-strategy.*

Proof. In the following, we call a synchronized set of consecutive players *maximal*, if the next players to both ends of the set play different strategies than the synchronized players. Obviously in every state S of Γ which is not an equilibrium the number of maximal, synchronized sets of players playing their 0-strategy equals the number of maximal, synchronized subsets of players playing their 1-strategy.

We now prove that in every maximal, synchronized set of consecutive players only the first player clockwise has an incentive to change her strategy. Thus, in every maximal set, there is only a single player who wants to change her strategy. Note that this suffices to prove the lemma.

First, consider a maximal, synchronized subset of consecutive players $\mathcal{N}' = \{i, \ldots j\}$ which all play their 0-strategy. Then the player $i - 1$ plays her 1-strategy, and therefore the players $i - 1$ and i share the resource r_i. In this case, player i can decrease her delay by changing to her 1-strategy. Other players $k \in \mathcal{N}'$, $k \neq i$, do not have an incentive to change their strategy as this would increase their delay.

Second, consider a maximal, synchronized subset of consecutive players $\mathcal{N}' = \{i, \ldots j\}$ which all play their 1-strategy. Then player $i - 1$ plays her 0-strategies and therefore no player currently allocates the resource r_i. Observe now that player i may decrease her delay by changing to her 0-strategy. Again, all other players $k \in \mathcal{N}'$, $k \neq i$, do not have an incentive to change their strategy as this would increase their delay. This is especially true for the last player, who currently allocates an overloaded resource. $\qquad\square$

Consider now the random best response schedule activating an unsatisfied player uniformly at random. From Lemma 12 we conclude that the total number of players playing their 0-strategy increases or decreases by 1 with probability $1/2$. Combining this observation with the observation that in a Nash equilibrium all players play the same strategy, we conclude that the random best response schedule is isomorphic to a random walk on a line with $n + 1$ vertices. Vertex v_i corresponds to the fact that i players play their 0-strategy. As the time of such a random walk to reach one of the two ends of the line is $O(n^2)$, the lemma follows.   □

4   Conclusions and Open Problems

In this paper, we presented a first step towards analyzing the convergence time in player-specific congestion games. We presented polynomial upper bounds on the convergence time in very simple games. The techniques we invented seem to be inapplicable to games on more general graphs. Therefore, we leave it as a challenging open question to bound the convergence time in more general games. However, we conjecture that a family of instances with exponential expected convergence time can be constructed.

Acknowledgment

The author wishes to thank Heiko Röglin and Berthold Vöcking for helpful discussions.

References

1. Ackermann, H., Röglin, H., Vöcking, B.: On the impact of combinatorial structure on congestion games. In: Proceedings 47th Annual IEEE Symposium on Foundations of Computer Science (FOCS), pp. 613–622 (2006)
2. Ackermann, H., Röglin, H., Vöcking, B.: Pure Nash equilibria in player-specific and weighted congestion games. In: Spirakis, P.G., Mavronicolas, M., Kontogiannis, S.C. (eds.) WINE 2006. LNCS, vol. 4286, pp. 50–61. Springer, Heidelberg (2006)
3. Berenbrink, P., Friedetzky, T., Goldberg, L.A., Goldberg, P., Hu, Z., Martin, R.: Distributed selfish load balancing. In: Proceedings 17th Annual ACM–SIAM Symposium on Discrete Algorithms (SODA), pp. 354–363. ACM, New York (2006)
4. Even-Dar, E., Kesselman, A., Mansour, Y.: Convergence time to Nash equilibria. In: Baeten, J.C.M., Lenstra, J.K., Parrow, J., Woeginger, G.J. (eds.) ICALP 2003. LNCS, vol. 2719, pp. 502–513. Springer, Heidelberg (2003)
5. Fabrikant, A., Papadimitriou, C., Talwar, K.: The complexity of pure Nash equilibria. In: Proceedings 36th Annual ACM Symposium on Theory of Computing (STOC), pp. 604–612 (2004)
6. Goldberg, P.W.: Bounds for the convergence rate of randomized local search in a multiplayer load-balancing game. In: Proceedings 23rd Ann. ACM SIGACT-SIGOPS Symposium on Principles of Distributed Computing (PODC), pp. 131–140 (2004)
7. Ieong, S., McGrew, R., Nudelman, E., Shoham, Y., Sun, Q.: Fast and compact: A simple class of congestion games. In: 20th National Conference on Artificial Intelligence (AAAI) (2005)
8. Lovász, L.: Random Walks on Graphs: A Survey. In: Combinatorics: Paul Erdős is Eighty. János Bolyai Mathematical Society, vol. 2, pp. 353–398 (1996)
9. Milchtaich, I.: Congestion games with player-specific payoff functions. Games and Economic Behavior 13(1), 111–124 (1996)
10. Rosenthal, R.W.: A class of games possessing pure-strategy Nash equilibria. International Journal of Game Theory 2, 65–67 (1973)

Communication Problems in Random Line-of-Sight Ad-Hoc Radio Networks*

Artur Czumaj[1] and Xin Wang[2]

[1] Department of Computer Science, University of Warwick, Coventry CV4 7AL, United Kingdom
czumaj@dcs.warwick.ac.uk
[2] Department of Computer Science, New Jersey Institute of Technology, Newark, NJ 07102-1982, USA
xw37@oak.njit.edu

Abstract. The *line-of-sight networks* is a network model introduced recently by Frieze et al. It considers wireless networks in which the underlying environment has a large number of obstacles and the communication can only take place between objects that are *close in space* and are *in the line of sight* to one another. To capture the main properties of this model, Frieze et al. proposed a new *random networks model* in which nodes are randomly placed on an $n \times n$ grid and a node can communicate with all the nodes that are in at most a certain fixed distance r and which are in the same row or column.

Frieze et al. concentrated their study on basic structural properties of the random line-of-sight networks and in this paper we focus on their communication aspects in the scenario of ad-hoc radio communication networks. We present efficient algorithms for two fundamental communication problems of *broadcasting* and *gossiping* in the classical ad-hoc radio communication model adjusted to random line-of-sight networks.

1 Introduction

In this paper we study basic communication properties of *random Line-of-Sight networks*, a model of wireless networks introduced recently by Frieze et al. [10]. The model of line-of-sight networks has been motivated by wireless networking applications in complex environments with obstacles. It considers scenarios of wireless networks in which the underlying environment has obstacles and the communication can only take place between objects that are close in space and are in the line of sight (are visible) to one another. In such scenarios, the classical random graph models [2] and random geometric network models [16] seem to be not well suited, since they do not capture main properties of environments with obstacles and of line-of-sight constraints. Therefore Frieze et al. [10] proposed a new random network model that incorporates two key parameters in such scenarios: range limitations and line-of-sight restrictions. In the model of Frieze

---

* Research supported in part by the *Centre for Discrete Mathematics and its Applications (DIMAP)*, University of Warwick, EPSRC grant EP/D063191/1.

J. Hromkovič et al. (Eds.): SAGA 2007, LNCS 4665, pp. 70–81, 2007.
© Springer-Verlag Berlin Heidelberg 2007

et al. [10], one places points randomly on a 2-dimensional grid and a node can see (can communicate with) all the nodes that are in at most a certain fixed distance and which are in the same row or column. One motivation is to consider urban areas, where the rows and the columns correspond to "streets" and "avenues" among a regularly spaced array of obstructions.

Frieze et al. [10] concentrated their study on basic structural properties of the line-of-sight networks like the connectivity, k-connectivity, etc. In this paper, we initiate the study of fundamental communication properties of the random line-of-sight networks in the scenario of *ad-hoc radio communication networks*. Our focus is on two classical communication problems: *broadcasting* and *gossiping*. In the broadcasting problem, a distinguished source node has a message that must be sent to all other nodes. In the gossiping problem, the goal is to disseminate the messages so that each node will receive messages from all other nodes.

1.1   Random Line-of-Sight Network and Communication Protocols

A *line-of-sight network* is defined on a set T of grid points $\{(x,y) : x,y \in \{1,2,\ldots,n\}\}$. Let p, called a *placement probability*, be a parameter of the network, $0 \le p \le 1$. Then for each lattice point x from T we place a wireless device at x independently at random with probability p.

To measure the distance between any lattice points we use the L_1-distance: for two points $p = (i,j)$ and $q = (i',j')$, we define $\mathsf{dist}(p,q) = |i - i'| + |j - j'|$. Each node has a (common) *range* r. There is a communication link between two nodes x, y iff x *and* y *are on the same straight line of the grid* and $\mathsf{dist}(x,y) \le r$.

As observed in [10], in the scenario of wireless communication, comparing with classic geometric network model, line-of-sight networks capture some important aspects of wireless networks. For example, in urban area, two wireless devices at different streets can not communicate each other even the Euclidean distance between them is quite small.

To study communication in the network, we consider an extension of the so-called *ad-hoc radio networks* model of communication [1,3,6,8,9,11,15]. We assume that all nodes have access to a global clock and work synchronously in discrete time steps called *rounds*. In radio networks the nodes communicate by sending messages through the edges of the network. In each round each node can either transmit the message to all its neighbors at once or can receive the message from one of its neighbors (be in the listening mode). A node x receives a message from its neighbor y in a given round iff: *(i)* x does not transmit (is in the listening mode) and *(ii)* y is the only neighbor of x that is transmitting in that round. In the case that constraint *(ii)* is violated, we say a *collision* occurs, in which case no message is received by x. In particular, we assume that x is unable to detect if a collision happened or none of its neighbors did transmit.

In the classical radio network model a node cannot receive any message if more than one of its neighbors transmits because the radio signals from these nodes will interfere. Although this definition is often used to model radio ad hoc networks, we believe that this definition of the collision here is too narrow to fit the framework of ad hoc networks in the line-of-sight model. Since in a radio

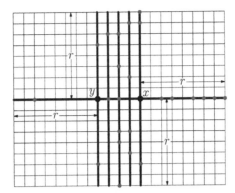

Fig. 1. The nodes on bold line could cause collision when y sends a message to x

network messages are sent out via microwave signals, signals can interfere each other if they overlap in space (rather than just by saying that when two or more neighbors transmit). For example, in Figure 1, a sending node z can interfere the node x from receiving the message of y, even though z is not neighbor of x.

To cope with this phenomenon, we add one more constraint to ensure that node x receives the message from y: (*iii*) no neighbor of y is transmitting *and* there is no node z that is transmitting and that lies on a grid-line perpendicular to the segment $\overline{xy}$ and is at distance at most r from the segment $\overline{xy}$, see, e.g., Figure 1. If either condition (*ii*) or (*iii*) is violated, then we say a collision occurs. It is easy to see that any gossiping algorithm that works in our model will certainly work in the traditional model, but not vice versa.

All protocols designed in this paper are distributed and there is no centralized coordinator. We assume the length of the message sent in each single round is at most polynomial in n, and thus, each node can combine multiple messages into one. Throughout this paper, we assume that each node has a unique integral ID that is bounded in polynomial of the number of nodes. The node only knows its own ID, the placement probability p, and its range r; the topology of the network is unknown for any node (this is the so-called *unknown network topology* model, see, e.g., [6,13]). In addition, we assume *in Section 3 only* that each node knows its own location in the grid. We do not use this assumption in Sections 4–5.

We say an algorithm *completes broadcasting in T rounds* if at the end of round T each node already received the message from the source. We say an algorithm *completes gossiping in T rounds* if at the end of round T each node already received messages from all other nodes.

1.2 Properties of Random Line-of-Sight Networks

Two parameters, r and p, play a critical role in the analysis of properties of the random line-of-sight networks. Frieze et al. [10] proved that when $r = o(n)$ and $r = \omega(\log n)$, if $r \cdot p = o(\log n)$, the network is disconnected with high probability, and therefore no full information exchange (including broadcasting and gossiping processes) can be performed in that case. Therefore we will assume that $r = o(n)$, $r = \omega(\log n)$, and $r \cdot p \geq c \cdot \log n$ for some sufficiently large constant c. This will ensure that the network is connected with high probability.

Assuming that $r \cdot p > c \cdot \log n$ for some sufficiently large constant c, we can also make some further assumptions about the structure of the input network. And so, it is easy to prove, that such a random line-of-sight network has minimum and maximum degree $\Theta(r\,p)$, and has *diameter* $D = \Theta(n/r)$, where these bounds hold with high probability. Besides these easy properties, we also need some other properties that are essential for our algorithms. Before we state them, let us introduce one more notation: for any $r \times r$ square in the grid T, we call the graph induced by the nodes in this sub-grid as an *r-graph*.

Lemma 1. [10] *If $rp > c\log n$ for certain constant c, then with high probability:*

(1) all r-graphs are connected, and
(2) the diameter of any r-graph is $\alpha = \Theta(\log r / \log(p\,r))$.

For simplicity of presentation, throughout the paper we will use term α as in the lemma above. All the claims in Lemma 1 hold *with high probability*, that is, with probability at least $1 - 1/n^3$. Therefore, from now on, we shall implicitly condition on these events.

1.3   Related Prior Works

Broadcasting and gossiping have been extensively studied in the ad-hoc radio networks model of communication. Before we present prior works on these problems, we want to clarify that in all these models, broadcasting and gossiping problems are considered on the *communication graph* induced by the accessibility of nodes: if the signal of node x can reach y, then there is an edge from x to y. Let us call the set of nodes that can interfere x from getting message from y a *collision set* (of x with respect to y), denoted by $c(x, y)$. In the classical radio network model studied before, $c(x, y)$ can be implicitly expressed by the communication graph: $c(x, y) = \{z : z \text{ is the neighbor of } y\}$. But in our model, there is no such a clean structure and collision sets $c(x, y)$ cannot be defined in term of the communication graph, see Section 2 for the definition. So all algorithms, even those working in general communication graphs, do not work without any modifications in our model. To the best of our knowledge, we have not seen any work on the broadcasting and gossiping problems in ad-hoc radio networks with collisions as defined in this paper.

Prior works with standard definition of collision sets. In the *centralized scenario*, when each node knows the entire network, Kowalski and Pelc [15] gave a centralized deterministic broadcasting algorithm running in $\mathcal{O}(D + \log^2 n)$ time and Gąsieniec et al. [11] designed a deterministic $\mathcal{O}(D + \Delta \log n)$-time gossiping algorithm, where D is the diameter and Δ the maximum degree of the network.

There has been also a very extensive research in the *non-centralized (distributed) setting in ad-hoc radio networks*, see, e.g., [3,6,12,14,15] and the references therein. In the model of *unknown topology (directed) networks*, there are known optimal randomized broadcasting $\mathcal{O}(D \log(n/D) + \log^2 n)$-time algorithms [6,14] and an almost optimal $\mathcal{O}(n \log^2 D)$-time deterministic algorithm [6]. The fastest randomized algorithm for gossiping in directed networks runs in $\mathcal{O}(n \log^2 n)$ time [6] and the fastest deterministic algorithm runs in $\mathcal{O}(n^{4/3} \log^4 n)$ time [12]. For undirected networks, both broadcasting and gossiping have deterministic $\mathcal{O}(n)$-time algorithms [1,4], what is asymptotically tight [1,13]. Clementi et al. [3] gave an algorithm that runs in $\mathcal{O}(D \Delta^2 \log n)$ time.

Dessmark and Pelc [8] consider broadcasting in ad-hoc radio networks in a model of *geometric networks* with the nodes knowing their own locations on the plane, and Elsässer and Gąsieniec [9], Chlebus et al. [5], and Czumaj and Wang [7] consider broadcasting in ad-hoc radio *random networks*.

1.4   New Contributions

In this paper we present the first thorough study of basic communication primitives in random line-of-sight networks.

We first consider in Section 3 the most powerful model in which each node knows its own geometric position in the grid (but it does not know positions of other nodes). In this model, we present a distributed deterministic algorithm that completes *gossiping* in $\mathcal{O}(r^4 p^2 \alpha \log n + n/r)$ steps, with α as in Lemma 1.

Next, we study the model in which each node knows its own ID, the values of n, r, and p, but it is not aware of any other information about the network. We believe that this is the main model to be studied in line-of-sight networks. We present two algorithms in this model. First, in Section 4, we design a distributed *deterministic* algorithm that completes *broadcasting* in $\mathcal{O}(r^4 p^2 \alpha \log n + n/r)$ steps. Next, in Section 5, we design a distributed *randomized* algorithm that completes *gossiping* in $\mathcal{O}(r^4 p^2 \alpha \log^2 n + n/r)$ steps. These two results demonstrate that even if only a very limited information about the network is known to the nodes, still broadcasting and gossiping can be performed very fast.

The running time of $\mathcal{O}(r^4 p^2 \alpha \log^2 n + n/r)$ steps for these algorithms may seem to be unimpressive, but this running time is especially efficient in the most interesting scenario when r is not too large and the product $r \cdot p$ is just a little larger than that needed to ensure the connectivity of the network. In particular, if $r \leq \mathcal{O}(n^{1/5}/\log^{3/5} n)$, our algorithms achieve the *optimal number of steps* which is proportional to the diameter of the network. Thus, we demonstrate that for a large range of the input parameters (including those being the most interesting) our algorithms are achieving asymptotically optimal running time.

Finally, in all the bounds above, we have assumed the model of collisions as discussed in Section 1.1. Still, our algorithms can be also run (without any modification) in the classical model of collisions in radio networks. Furthermore, it is not difficult to see that in that case one can speed up all our algorithms to remove a factor $\Theta(r^2)$ from the first term in the running times. And so, the first two algorithms have the running time of $\mathcal{O}(r^2 p^2 \alpha \log n + n/r)$ and the third one runs in $\mathcal{O}(r^2 p^2 \alpha \log^2 n + n/r)$ time. This yields an optimal number of steps for $r = \mathcal{O}(n^{1/3}/\log n)$. The details are deferred to the full version of the paper.

2   Preliminaries

Let V be the set of nodes in the grid. For any node x, define $N(x)$ to be the set of nodes are reachable from x in one hop, $N(x) = \{y \in V : \text{dist}(v, u) \leq r$ and x, y are on the same straight line$\}$, where $\text{dist}(v, u)$ is the distance between v and u. Any node in $N(v)$ is called a *neighbor* of v, and set $N(v)$ is called the *neighborhood* of v. For any $X \subseteq V$, let $N(X) = \bigcup_{x \in X} N(x)$. Define the *kth neighborhood* of a node v, $N^k(v)$, recursively as follows: $N^0(v) = v$ and $N^k(v) = N(N^{k-1}(v))$ for $k \geq 1$. Let Δ be the *maximum degree* and D be the *diameter* of the radio network. As we mentioned earlier, in our model $\Delta = \Theta(r\, p)$ and $D = \Theta(n/r)$, with high probability.

Definition 1 (Collision sets). *Let $x, y \in T$ with $x \in N(y)$. We define the collision set for the communication from y to x, denoted by $C(y, x)$, to be the set of nodes that can interfere x from receiving a message from y. Set $C(y, x)$ contains all nodes $z \in T$ that satisfy one of the following:*

1. *$z \in N(x) \cup N(y)$, or*
2. *there is a grid point q such that (i) q lies on the segment connecting x and y, (ii) grid line $\overline{zq}$ is orthogonal to the grid line $\overline{xy}$, and (iii) $\text{dist}(z, q) \leq r$.*

It is easy to see that $C(y, x) = \mathcal{O}(r^2 p)$ with high probability.

Strongly selective families. Let k and m be two arbitrary positive integers with $k \leq m$. Following [3], a family $\mathcal{F}$ of subsets of $\{1, \ldots, m\}$ is called (m, k)-*strongly-selective* if for every subset $X \subseteq \{1, \ldots, m\}$ with $|X| \leq k$, for every $x \in X$ there exists a set $F \in \mathcal{F}$ such that $X \cap F = \{x\}$. It is known (see, e.g., [3]) that for every k and m, there exists a (m, k)-strongly-selective family of size $\mathcal{O}(k^2 \log m)$.

In the last years the concept of strongly-selective families has been successfully used to design fast deterministic distributed broadcasting and gossiping algorithms. In particular, Clementi et al. [3] used this approach to obtain a deterministic distributed gossiping algorithm for general radio networks (with the standard notion of collision sets) that runs in $\mathcal{O}(D\, \Delta^2 \log n)$ time.

We can use this approach for line-of-sight networks (and for our notion of collision sets) to obtain the following result.

Lemma 2. *In random line-of-sight networks, for any integer k, in (**deterministic**) time $\mathcal{O}(k\, r^4 p^2 \log n)$ all nodes can send their messages to all nodes in their kth neighborhood. The algorithm may fail with probability at most $1/n^2$ (where the probability is wrt. the random choice of the nodes in the network).*

Proof. Our arguments follow a nowadays standard approach of applying selective families to broadcasting and gossiping in radio ad-hoc networks, see, e.g., [3].

Since in our setting, for each pair of nodes x and y, we have a collision set of size $|C(y, x)| = \mathcal{O}(r^2 p)$, with high probability.

Let us consider a random line-of-sight network with at most n^2 nodes. Assume, wlog that all IDs are distinct integers in $\{1, 2, \ldots, n^\lambda\}$, for a constant λ. Let $\mathcal{F} = \{F_1, F_2, \ldots\}$ be an $(n^\lambda, \Theta(r^2 p))$-strongly-selective family of size $\mathcal{O}(r^4 p^2 \log n)$; the existence of such family follows from our discussion above. Then, consider a protocol in which in step t only the nodes whose IDs are in the set F_t transmit. By the strong selectivity property, for every node u and its every neighbor v there is at least one time step when u does not transmit and v is the only node of $C(v, u)$ that transmits in that step since $C(v, u) = \mathcal{O}(r^2 p)$. Therefore, every node will receive a message from all its neighbors after $\mathcal{O}(r^4 p^2 \log n)$ steps. Hence, we can repeat this procedure to ensure that after $\mathcal{O}(k r^4 p^2 \log n)$ steps, every node will receive a message from its entire kth neighborhood.   □

The following is an immediate corollary of Lemma 2 obtained by setting $k = D$ (which corresponds to the bound from [3] in our setting):

Corollary 1. *Distributed gossiping in random line-of-sight networks can be performed in* **deterministic** *time* $\mathcal{O}(D r^4 p^2 \log n) = \mathcal{O}(n r^3 p^2 \log n)$. *The algorithm may fail with probability at most* $1/n^2$.

Since $r p = \Omega(\log n)$, the running time of this algorithm is in the best case $\Omega(n \log^4 n)$, and thus it is *superlinear*. The goal of this paper is to develop algorithms that are faster, optimally, those that achieve the running time $\mathcal{O}(D)$, which is a trivial asymptotic lower bound for broadcasting and gossiping.

3   Deterministic Algorithm with Position Information

We consider the gossiping problem in random line-of-sight networks in the model, where each node knows its own geometric position in the grid. In such model, Dessmark and Pelc [8] give a deterministic distributed broadcasting algorithm that runs in $\mathcal{O}(D)$ time. It can be applied to solve the broadcasting problem in our model, with the same running time. We can prove a similar result for gossiping by extending the preprocessing phase from [8] and use an appropriate strongly-selective family to collect information about the neighbors of each node.

Theorem 1. *If every input node knows its location in the* $n \times n$ *grid, then the algorithm Gossiping-Known-Locations-D below will complete gossiping in a random line-of-sight network in deterministic time* $\mathcal{O}(\alpha r^4 p^2 \log n + n/r)$. *The algorithm may fail with probability at most* $1/n^2$.

Let us introduce some notations. We label the horizontal lines in the grid as $H_1, \ldots, H_n$ from bottom to top; label the vertical lines in the grid as $V_1, \ldots, V_n$ from left to right. We refer the crossing point of H_i and V_j as (H_i, V_j). For each H_i, we further divide H_i into *segments* of length of $r/2$, except for the last segment which is of length at most $r/2$, and label them as: $H_{i,1}, \ldots, H_{i,\lceil 2n/r \rceil}$ from left to right. Define $V_{j,1}, \ldots, V_{j,\lceil 2n/r \rceil}$ in a similar way, from bottom to top.

Gossiping-Known-Locations-D

Preprocessing: do *local gossiping* using Lemma 2 (with $k = \alpha$)
    to ensure that each node knows messages and positions of its neighbors
  for j = 1 **to** $\lceil 2n/r \rceil$ **do:**
    for c = 1 **to** 4 **do:**
      for each i with $i \bmod 4 + 1 \equiv c$ in parallel **do:**
        the node with the minimum ID in the $H_{i,j}$ transmits
  for j = $\lceil 2n/r \rceil$ **downto** 1 **do:**
    for c = 1 **to** 4 **do:**
      for each i with $i \bmod 4 + 1 \equiv c$ in parallel **do:**
        the node with the minimum ID in the $H_{i,j}$ transmits
  for j = 1 **to** $\lceil 2n/r \rceil$ **do:**
    for c = 1 **to** 4 **do:**
      for each i with $i \bmod 4 + 1 \equiv c$ in parallel **do:**
        the node with the minimum ID in the $V_{i,j}$ transmits
  for j = $\lceil 2n/r \rceil$ **downto** 1 **do:**
    for c = 1 **to** 4 **do:**
      for each i with $i \bmod 4 + 1 \equiv c$ in parallel **do:**
        the node with the minimum ID in the $V_{i,j}$ transmits
Postprocessing: do *local gossiping* using Lemma 2 (with $k = \alpha$)

It is easy to see that for $r\,p \geq c \log n$ with a sufficiently large c, with high probability, there is a node in each segment. For any node x, we define $M_t(x)$ as the messages known by x at step t. For any segment S, we define $M_t(S)$ as the common messages known by all nodes in segment S at step t.

Proof. After the preprocessing in the algorithm Gossiping-Known-Locations-D, by Lemma 2, every node knows which segment it belongs to, and every node knows all other nodes in its segment, including their messages and positions. Therefore, in every segment, all the nodes from that block can select a single representative who will be the only node transmitting (for the entire segment) in all the following time slots.

Let us call all nodes that are scheduled to send by the algorithm *representative nodes*. By our definition of the segment, in any time slot, when a segment (its representative) sends a message, the nearest sending segment is at distance of $2\,r$ from it. So after each sending, the sending segment, say $H_{i,j}$, will successfully send its message $M_t(H_{i,j})$ to segments $H_{i,j-1}$ and $H_{i,j+1}$ if there are such segments. The statement is also true for any segment $V_{i,j}$. For any two nodes v and u in the grid, the algorithm will sent the message of u to v successfully: There are two representative nodes that are within the $r \times r$ square centered at u and v respectively, with high probability. After preprocessing, the message of u will be sent to its representative node. Then after $\mathcal{O}(n/r)$ steps, the message of u will be sent to the representative node of v. After the postprocessing, the message of u will eventually reach v.

By Lemma 2, the running time of the preprocessing is $\mathcal{O}(\alpha\,r^4\,p^2 \log n)$, and so is the running time of the postprocessing phase. Therefore the total running time of Gossiping-Known-Locations-D is $\mathcal{O}(\alpha\,r^4\,p^2 \log n + n/r)$. $\qquad\square$

4   Broadcasting and Deterministic Gossiping with a Leader

We now move to a more natural model in which each node knows the values of n, r, and p, knows its own ID (which is a unique integer bounded by a polynomial of n), but it is not aware of any other information about the network. In particular, the node does not know its own location. We believe that this is the main model for the study of principles of the communication in random line-of-sight networks.

We begin our discussion with a slightly relaxed model: there is a special node (*leader*) ℓ in the network, such that ℓ knows that she is the leader, and all other nodes in the network know that they are not the leader. We will show that in this model distributed gossiping can be done *deterministically* in time $\mathcal{O}(\alpha\,r^4\,p^2\,\log n + n/r)$. This will immediately imply a deterministic distributed broadcasting algorithm with asymptotically the same running time.

We start with the same preprocessing as that used in algorithm Gossiping-Known-Locations-D from Section 3. This takes $\mathcal{O}(\alpha\,r^4\,p^2\,\log n)$ steps. After the preprocessing, each node knows its second neighborhood with high probability.

4.1   Gossiping Along a Grid Line

We first consider the gossiping among the nodes belonging to the same grid line.

We begin with two lemmas that estimate the size of the join neighborhood in random line-of-sight networks. The lemmas follow easily from Chernoff bounds.

Lemma 3. *For any u, v belonging to the same grid line, with high probability:*

(i) if $\mathsf{dist}(u,v) \leq r/2$ then $|N(u) \cap N(v)| \geq 1.3\,r\,p$, and
(ii) if $\mathsf{dist}(u,v) \geq r$ then $|N(u) \cap N(v)| \leq 1.2\,r\,p$.

Lemma 4. *For any node u, if the distance between u and the nearest boundary is greater than or equal to r, then, in each of four directions, with high probability, there is a neighboring node v of u such that $1.2\,r\,p \leq |N(u) \cap N(v)| \leq 1.3\,r\,p$.*

The process of the gossiping among the nodes on a grid-line is initialized by one specific node (call it a *launching node*). The launching node u checks its second neighborhood $N^2(u)$, and selects one *representative node*, say v, such that $1.2\,r\,p \leq |N(u) \cap N(v)| \leq 1.3\,r\,p$. Then, u sends a message to v with the aims: *(i)* u transmits its message to v and *(ii)* u informs v that it is picked as representative node. Because of Lemmas 3 and 4, we know $r/2 \leq \mathsf{dist}(u,v) \leq r$.

The process of gossiping along a grid line is working in steps. At the beginning of each step, a node ϖ_t receives a message from node ϖ_{t-1}, and ϖ_t is informed that it is the representative node. Then ϖ_t will pick a representative node ϖ_{t+1} for the next step, and then send a message to ϖ_{t+1} to inform about it. ϖ_t picks ϖ_{t+1} by checking $N^2(\varpi_t)$, and selecting as ϖ_{t+1} any node fulfilling: *(i)* $1.2\,r\,p \leq |N(\varpi_t) \cap N(\varpi_{t+1})| \leq 1.3\,r\,p$, *(ii)* $|N(\varpi_{t-1}) \cap N(\varpi_{t+1})| \leq 1.3\,r\,p$.

Because of Lemmas 3 and 4, it is easy to see that $r/2 \leq \mathsf{dist}(\varpi_{t+1}, \varpi_t) \leq r$. Moreover, ϖ_{t+1} and ϖ_{t-1} are at the different sides of ϖ_t, for otherwise

$\mathsf{dist}(\varpi_{t+1}, \varpi_{t-1}) \leq r/2$ and then the second constraint would be violated with high probability. If in one step ϖ_t is unable to find the ϖ_{t+1} as defined above, then the distance between ϖ_t and its nearest boundary is less than r. In that case ϖ_t simply stops the process and makes itself as the *last representative node*.

We run this process for $2\,n/r$ steps and call these steps *Phase 1*. Then the last representative node, say ϖ_{t+1}, picks ϖ_t as next representative node and initialize the process again, for another $2\,n/r$ steps. These steps define *Phase 2*. Then, the representative nodes in Phase 2 send their messages in reverse order and with that the gossiping along straight-line will be done. These steps form *Phase 3*. The total running time is $\mathcal{O}(n/r)$.

4.2   Broadcasting and Gossiping with the Leader in the Whole Grid

Now, we are ready to present our gossiping algorithm in the model with a distinguished leader ℓ. First, the leader will pick an arbitrary direction and do gossiping along the corresponding grid line. As a by product of this algorithm, a set of representative nodes will be chosen and the minimum distance between any pair of them is greater than or equal to $r/2$. Next, each of the representative nodes treat itself as the pseudo-leader, and do gossiping along a grid line in parallel, in an orthogonal direction to that first chosen by the leader. There are two issues to be solved. First, since the distance between these pseudo leaders could be as small as $r/2$, we need to interleave the transmissions in adjacent pseudo-leaders, which yields a constant-factor slow-down. Second, by checking its second neighborhood, a pseudo-leader can indeed find a node in orthogonal direction that first chosen by the leader. (ϖ_i can pick a node y from its neighbors such that $1.2\,r\,p \leq |N(\varpi_t) \cap N(y)| \leq 1.3\,r\,p$ and $|N(\varpi_{t-1}) \cap N(y)| \equiv 1$.)

Next, we repeat the whole process once again. It is easy to see that gossiping is done among all representative nodes. For any pair of nodes u and v, there are two representative nodes that are within the $r \times r$ squares centered at u and v respectively, with high probability. After preprocessing, u will send its message to its representative nodes. The message of u then will be sent to the representative node of v in the following steps. Let us run the postprocessing defined in the algorithm of the Section 3, the message of u will be sent to v. The running time is $\mathcal{O}(\alpha\,r^4\,p^2\,\log n + n/r)$.

Theorem 2. *If there is a leader in the random line-of-sight network, then gossiping can be completed in **deterministic** time $\mathcal{O}(\alpha\,r^4\,p^2\,\log n + n/r)$. The algorithm may fail with probability at most $1/n^2$.*

Theorem 2 immediately implies the following result for broadcasting.

Theorem 3. *Distributed broadcasting in random line-of-sight networks can be performed in deterministic time $\mathcal{O}(\alpha\,r^4\,p^2\,\log n + n/r)$.*

5   Fast Distributed Randomized Gossiping

We continue our study of the model of random line-of-sight networks in which each node knows the values of n, r, and p, knows its own ID, but it is not aware of any other information about the network. As we have seen in the previous section, if the nodes in the network can elect a leader then gossiping could be done within the time bounds stated in Theorem 2. In this section, we will show that randomized leader election can be solved efficiently.

At the beginning, each node independently and uniformly at random selects itself as the *leader* with probability $\frac{\log n}{n^2 p}$. By simple probabilistic arguments, one can prove that exactly $\Theta(\log n)$ leaders are chosen with high probability. Then we want to execute a distributed *minimum finding* algorithm to eliminate all of them but one, and the one chosen will have the lowest ID.

Each of these leaders will pick four representative nodes along four directions, respectively, and execute the process of *Phase 1* as described in Section 4.1. Let us call it *fast transmission*. We interleave fast transmission with the preprocessing of the algorithm in Section 3. Let us call it *slow transmission*. In the odd steps, every node follows the schedule of fast transmission, and in the even steps, every node follow the schedule of slow transmission. Since we have more than one leader, it is possible that transmission *collisions* can occur. We are able to detect these collisions, because after ϖ_t picks ϖ_{t+1} as the next representative node and informs it, in the next step, ϖ_t is expected to receive an acknowledgement of the successful transmission from ϖ_{t+1}. If the acknowledgement is not received, ϖ_t knows that a collision happened.

When a node, say u, detects a collision, in the following $\mathcal{O}(\alpha\, r^4\, p^2\, \log n)$ steps u will send nothing (stay in the listening mode) in odd steps, and run slow transmission as before in even steps. By the property of the strongly selective family, u will eventually receive the messages of other representative nodes that are transmitting for their own leaders during this period. Then u compares the ID of its own leader with all other leader's ID that it just received. If (at least) one of those ID is smaller than the ID of its leader, u will send an "eliminating" message back to its leader, reversely along the path through which the leader sent the message to it. The leader eliminates itself after getting this message. If the ID of u's leader is the smallest one, u resumes the fast transmission. Altogether, a node will encounter at most $\mathcal{O}(\log n)$ collisions when it transmits toward any boundary. Therefore the slow down caused by collisions is small and the total running time is $\mathcal{O}(r^4\, p^2\, \alpha\, \log^2 n + n/r)$.

Now we can see that the leader is either eliminated or successfully transmits its ID along four directions. Thus, for any pair of surviving leaders, there is a pair of representative nodes in one $r \times r$ square. If we run the postprocessing from Section 3, the two representative nodes will exchange information about their leaders. Again, the representative node that holds the larger ID will transmit an "eliminating" message to its leader. The running time is $\mathcal{O}(r^4\, p^2\, \alpha\, \log^2 n + n/r)$. After this procedure, all but one leader with the smallest ID will survive.

Finally, we can combine our analysis above with Theorem 2, to obtain:

Theorem 4. *In the random line-of-sight network, distributed gossiping can be completed in randomized time* $\mathcal{O}(r^4\, p^2\, \alpha\, \log^2 n + n/r)$, *with high probability.*

6   Conclusions

We have presented three efficient algorithms for broadcasting and gossiping in the model of random sight-of-line networks. If $r = \mathcal{O}(n^{1/5}/\log^{3/5} n)$, then all our algorithms perform the optimal number of $\mathcal{O}(D)$ steps. While it is very interesting to extend the optimality of these bounds to larger values of r, we believe that the case $r = \mathcal{O}(n^{1/5}/\log^{3/5} n)$ covers the most interesting cases, when the graph is relatively sparse and each node is able to communicate only with the nodes that are not a large distance apart. Therefore, in our opinion the most interesting specific open problem left in this paper is to extend the result from Section 5 to obtain a distributed *deterministic* algorithm for gossiping. Even more interesting is a more general question: what are important aspects of random sight-of-line networks to perform fast communication in these networks. Our work is only the very first step in that direction.

References

1. Bar-Yehuda, R., Goldreich, O., Itai, A.: On the time-complexity of broadcast in multi-hop radio networks: An exponential gap between determinism and randomization. JCSS 45(1), 104–126 (1992)
2. Bollobás, B.: Random Graphs, 2nd edn. Academic Press, London (1985)
3. Clementi, A.E.F., Monti, A., Silvestri, R.: Distributed broadcast in radio networks of unknown topology. TCS, 302, 337–364 (2003)
4. Chlebus, B.S., Gąsieniec, L., Gibbons, A., Pelc, A., Rytter, W.: Deterministic broadcasting in ad hoc radio networks. Distrib. Comput. 15(1), 27–38 (2002)
5. Chlebus, B.S., Kowalski, D.R., Rokicki, M.A.: Average-time coplexity of gossiping in radio networks. In: Flocchini, P., Gasieniec, L. (eds.) SIROCCO 2006. LNCS, vol. 4056, pp. 253–267. Springer, Heidelberg (2006)
6. Czumaj, A., Rytter, W.: Broadcasting algorithms in radio networks with unknown topology. Proc. 44th FOCS, pp. 492–501 (2003)
7. Czumaj, A., Wang, X.: Gossiping in random geometric ad-hoc radio networks. Manuscript (2007)
8. Dessmark, A., Pelc, A.: Broadcasting in geometric radio networks. Proc.13th SPAA, pp. 59–66 (2001)
9. Elsässer, R., Gąsieniec, L.: Radio communication in random graphs. JCSS 72, 490–506 (2006)
10. Frieze, A., Kleinberg, J., Ravi, R., Debany, W.: Line-of-sight networks. In: Proc. 18th SODA, pp. 968–977 (2007)
11. Gąsieniec, L., Peleg, D., Xin, Q.: Faster communication in known topology radio networks. In: Proc. 24th PODC, pp. 129–137 (2005)
12. Gąsieniec, L., Radzik, T., Xin, Q.: Faster deterministic gossiping in directed ad-hoc radio networks. In: Proc. 9th SWAT, pp. 397–407 (2004)
13. Kowalski, D., Pelc, A.: Deterministic broadcasting time in radio networks of unknown topology. In: Proc. 43rd FOCS, pp. 63–72 (2002)
14. Kowalski, D., Pelc, A.: Broadcasting in undirected ad hoc radio networks. Distrib. Comput. 18, 43–57 (2005)
15. Kowalski, D., Pelc, A.: Optimal deterministic broadcasting in known topology radio networks. Distrib. Comput. 19(3), 185–195 (2007)
16. Penrose, M.D.: Random Geometric Graphs. Oxford University Press, Uk (2003)

Approximate Discovery of Random Graphs*

Thomas Erlebach[1], Alexander Hall[2], and Matúš Mihaľák[3]

[1] Department of Computer Science, University of Leicester, University Road,
Leicester LE1 7RH, UK
te17@mcs.le.ac.uk
[2] Department EECS, UC Berkeley, CA 94720, USA
alex.hall@gmail.com
[3] Institute for TCS, ETH Zurich, CH-8092 Zurich, Switzerland
matus.mihalak@inf.ethz.ch

Abstract. In the layered-graph query model of network discovery, a query at a node v of an undirected graph G discovers all edges and non-edges whose endpoints have different distance from v. We study the number of queries at randomly selected nodes that are needed for approximate network discovery in Erdős-Rényi random graphs $G_{n,p}$. We show that a constant number of queries is sufficient if p is a constant, while $\Omega(n^\alpha)$ queries are needed if $p = n^\varepsilon/n$, for arbitrarily small choices of $\varepsilon = 3/(6 \cdot i + 5)$ with $i \in \mathbb{N}$. Note that $\alpha > 0$ is a constant depending only on ε. Our proof of the latter result yields also a somewhat surprising result on pairwise distances in random graphs which may be of independent interest: We show that for a random graph $G_{n,p}$ with $p = n^\varepsilon/n$, for arbitrarily small choices of $\varepsilon > 0$ as above, in any constant cardinality subset of the nodes the pairwise distances are all identical with high probability.

1 Introduction

A fundamental problem in the study of complex networks is how to obtain accurate information about the topology of a network using a limited number of measurements or observations. For example, attempts to map the Internet can be based on traceroute experiments [1] or on the analysis of BGP routing tables [2]. A simplified theoretical model of such *network discovery* settings, the so-called *layered-graph query model*, has been introduced in [3]. The goal is to discover the edges and non-edges (for $u, v \in V$, we call $\{u, v\}$ a non-edge if it is not an edge of the graph) of an unknown graph or network $G = (V, E)$ using a minimum number of queries; a query at a node v reveals all edges and non-edges whose endpoints have different distance from v.

The layered-graph query model can be interpreted in the following way: A query at v yields the shortest-path subgraph rooted at v, i.e., the set of all edges

---

* Work partially supported by European Commission - Fet Open project DELIS IST-001907 Dynamically Evolving Large Scale Information Systems, for which funding in Switzerland is provided by SBF grant 03.0378-1.

J. Hromkovič et al. (Eds.): SAGA 2007, LNCS 4665, pp. 82–92, 2007.
© Springer-Verlag Berlin Heidelberg 2007

on shortest paths between v and any other node. To see that this is equivalent to our definition (where a query yields all edges and non-edges between vertices of different distance from v), note that an edge connects two vertices of different distance from v if and only if it lies on a shortest path between v and one of these two vertices. Furthermore, the shortest-path subgraph rooted at v implicitly confirms the absence of all edges between vertices of different distance from v that are not part of the shortest-path subgraph.

This model clearly is an abstraction of reality. Two real-life scenarios where the results of queries come close to yielding shortest-path subgraphs are: traceroute based experiments (done, e.g., by the DIMES project [1]) and querying border gateway protocol routers (pursued, e.g., by the RouteViews project [2]). In a recent paper [4] several snapshots of the Internet graph obtained by both approaches are compared. In particular, it is checked how well the layered-graph model fits to the actually collected data.

In the off-line version of network discovery, the goal is to verify with as few queries as possible a given graph or network $G = (V, E)$. In this case we also speak of *network verification*.

Simulation experiments reported in [5] with (scale-free as well as Erdős-Rényi) random graphs indicate that the number of queries needed to discover all edges and non-edges typically grows with the size of the graph, as expected, but in some cases appears to be bounded by a small constant independent of the size of the graph if only a large fraction (say, 95%) of the edges and of the non-edges needs to be discovered. This shows that for the practically relevant goal of approximate network discovery, a surprisingly small number of queries is often sufficient. Motivated by this experimental result, we now study this phenomenon analytically for Erdős-Rényi random graphs $G_{n,p}$. These are graphs on n nodes in which each possible edge is present independently with probability p. We consider the simple query strategy that selects the query nodes uniformly at random. We say that a set of random queries approximately discovers $G_{n,p}$ in expectation, if the expected number of edges discovered by the queries is at least a ρ-fraction of all edges, and the analogous condition is satisfied for non-edges. Here, ρ is a constant such as 0.95.

Surprisingly, we find that if p is a constant strictly between 0 and 1 (i.e., if we consider dense $G_{n,p}$ graphs), then a constant number of query nodes is sufficient to approximately discover $G_{n,p}$ in expectation, but if $p = n^\varepsilon/n$, for an arbitrarily small constant $\varepsilon = 3/(6 \cdot i + 5)$ with $i \in \mathbb{N}$, then $\Omega(n^\alpha)$ queries are necessary, where $\alpha > 0$ is a constant depending on ε. Our results show that the number of random queries needed to approximately discover $G_{n,p}$ depends on the density of the graph, and in the query model considered it is actually easier to discover dense random graphs than relatively sparse ones.

The results of this paper are mainly of theoretical interest and can be seen as first steps. We believe it would be of interest to do a similar analysis for scale free (e.g., Barabási-Albert [6]) random graphs, since they more realistically capture properties of the Internet graph.

Related Work. There are several ongoing large-scale efforts to collect data representing local views of the Internet. Here we will only mention two. The most prominent one is probably the RouteViews project [2] by the University of Oregon. It collects data from a large number of so-called border gateway protocol routers. Essentially, for each router—which can be seen as a node in the Internet graph on the level of autonomous systems—its list of paths (to all other nodes in the network) is retrieved. More recently and, due to good publicity, very successfully, the DIMES project [1] has started collecting data with the help of a volunteer community. Users can download a client that collects paths in the Internet by executing successive traceroute commands. A central server can direct each client individually by specifying which routes to investigate. Data obtained by these or similar projects has been used in heuristics to obtain maps of the Internet, basically by simply overlaying possible paths found by the respective project. There is an extensive body of related work studying various aspects of this approach, see, e.g., [1,2,7,8,9,10,11,12,13,14,15,16].

In [3,5], the network discovery and verification problems are introduced and several results for the layered-graph query model are presented. It is shown that the network verification problem cannot be approximated within a factor of $o(\log n)$ unless $\mathcal{P} = \mathcal{NP}$, proving that an approximation algorithm from [17] (see below) is best possible, up to constant factors. A useful lower bound formula is given for the optimal number of queries of a graph. A discussion of simulation experiments for four different heuristic discovery strategies on various types of graphs, including several random graph models, can be found in [5]. Moreover, the on-line setting (network discovery) is studied and several lower and upper bounds on the competitive ratio are given. A number of results for both the on-line and off-line setting have also been derived for the much weaker *distance query model* [18,5], in which a query at node v reveals only the distances to all other nodes.

It turns out that the network verification problem in the layered-graph query model has previously been considered as the problem of placing landmarks in graphs [17]. Here, the motivation is to place landmarks in as few vertices of the graph as possible in such a way that each vertex of the graph is uniquely identified by the vector of its distances to the landmarks. The smallest number of landmarks that are required for a given graph G is also called the *metric dimension* of G [19]. For a survey of known results, we refer to [20].

The problem of determining whether k landmarks suffice (i.e., of determining if the metric dimension is at most k) is long known to be $\mathcal{NP}$-complete [21]; the mentioned inapproximability of $o(\log n)$ [3] for the network verification problem transfers directly to the problem of minimizing the number of landmarks. In [17] it is shown that the problem admits an $O(\log n)$-approximation algorithm based on SETCOVER. For trees, they show that the problem can be solved optimally in polynomial time. Furthermore, they prove that one landmark is sufficient if and only if G is a path, and discuss properties of graphs for which 2 landmarks suffice. They also show that if k landmarks suffice for a graph with n vertices and diameter D, we must have $n \leq D^k + k$. For d-dimensional hypercubes, it was shown in [22] (using an earlier result from [23] on a coin weighing problem) that the metric dimension is asymptotically equal to $2d/\log_2 d$. See [24] for further results on the metric dimension of Cartesian products of graphs.

Our Contribution and Outline. In Section 2 we give some preliminary definitions concerning (random) graphs and the layered-graph query model of network discovery. The stated results in $G_{n,p}$ graphs are presented in Section 3. Our analysis for constant p in Section 3.1 is based on the observation that the probability that a query at node q discovers an edge or non-edge $\{u, v\}$ is at least $2p(1 - p)$, which is the probability that q is adjacent to one of u, v but not the other.

For the case of $p = n^\varepsilon/n$, treated in Section 3.2, we use bounds from [25] on the size of the i-neighborhood and on the size of the i-th breadth-first search layer of a node in $G_{n,p}$, for arbitrarily small $\varepsilon = 3/(6 \cdot i + 5)$ depending on the choice of i. These bounds allow us to show that for an edge or non-edge $\{u, v\}$, a query node q is very likely to have the same distance from u and v (and thus does not discover the edge or non-edge). We generalize this in Section 3.3 to obtain the following result: For a random graph $G_{n,p}$ with $p = n^\varepsilon/n$, for arbitrarily small choices of $\varepsilon > 0$ as above, in any constant cardinality subset of the nodes the pairwise distances are all identical, with high probability (w.h.p.).

2   Preliminaries

Graphs and Neighborhoods. With $G = (V, E)$ we denote an undirected graph with $|V| = n$ nodes. For two distinct nodes $u, v \in V$, we say that $\{u, v\}$ is an *edge* if $\{u, v\} \in E$ and a *non-edge* if $\{u, v\} \notin E$. The set of non-edges of G is denoted by $\overline{E}$. For $u, v \in V$, let $d(u, v)$ be the distance between the nodes u, v, i.e., the number of edges on a shortest path between u and v. For a graph G and a node $v \in V$, the set of nodes at distance i of v is denoted as the *i-th layer*: $\Gamma_i(v) = \{u \in V | d(v, u) = i\}$. We define the *$i$-neighborhood* $N_i(v) = \bigcup_{j=0}^{i} \Gamma_j(v)$ to be the set of nodes within distance i of v.

$G_{n,p}$ denotes an Erdős-Rényi random graph on n nodes in which a pair of nodes appears as an edge with probability p.

The Layered-Graph Query Model. A *query* is specified by a node $v \in V$ and is called a query *at* v or simply the query v. The answer of a query at v consists of a set E_v of edges and a set $\overline{E}_v$ of non-edges. These sets are determined as follows. Let E_v be the set of all edges connecting vertices in different layers (from v), and $\overline{E}_v$ be the set of all non-edges whose endpoints are in different layers. Because the query result can be seen as a layered graph, we refer to this query model as the *layered-graph query model*. We say a query $v \in V$ *discovers a node pair* $u, w \in V$ if u, w are in different layers (from v), i.e., if $d(v, u) \neq d(v, w)$ holds.

A set $Q \subseteq V$ of queries discovers (all edges and non-edges of) a graph $G = (V, E)$, if $\bigcup_{q \in Q} E_q = E$ and $\bigcup_{q \in Q} \overline{E}_q = \overline{E}$. In the off-line case, we also say "verifies" instead of "discovers". The network verification problem is to compute, for a given network G, a smallest set of queries that verifies G. The network discovery problem is the on-line version of the network verification problem. Its goal is to compute a smallest set of queries that discovers G. Here, the edges and non-edges of G are initially unknown to the algorithm, the queries are made sequentially, and the next query must always be determined based only on the answers of previous queries.

Discovering a Large Fraction of a Graph. Let $\rho \in (0,1]$ be a constant, typically a "large" value close to 1. We say a query set $Q \subseteq V$ discovers a ρ-fraction of the graph, if $|\bigcup_{q \in Q} E_q| \geq \rho \cdot |E|$ and $|\bigcup_{q \in Q} \overline{E}_q| \geq \rho \cdot |\overline{E}|$.

Note that we require separately that a fraction of all edges and that a fraction of all non-edges should be discovered. This is important, since another seemingly natural definition which requires only that a fraction of all node pairs should be discovered, might be misleading for the interesting case of sparse graphs. Here a query set discovering almost all non-edges but only some of the edges would be a valid solution, since the number of edges is small compared to the total number of node pairs. However, since only few edges were discovered, the resulting graph is far away from the actual one. This is avoided by the separate treatment of edges and non-edges.

For a random graph or if Q is a random variable, we say Q discovers a ρ-fraction of the graph in expectation, if

$$\mathbb{E}\left[|\cup_{q \in Q} E_q|\right] \geq \rho \cdot \mathbb{E}\left[|E|\right] \quad \text{and} \quad \mathbb{E}\left[|\cup_{q \in Q} \overline{E}_q|\right] \geq \rho \cdot \mathbb{E}\left[|\overline{E}|\right].$$

3   Discovering a Large Fraction of a Random Graph

In this section we study the discovery strategy RANDOM which simply picks a given number k of query nodes at random from V (using the uniform distribution). We show that in a random graph $G_{n,p}$ already a constant number of such queries suffices to discover a ρ-fraction of the graph in expectation, if p is a constant.

Since one of the main motivations for studying the network discovery setting is to discover the Internet graph, the case of sparse graphs is practically more relevant. Interestingly, if $p = n^\varepsilon/n$ for certain arbitrarily small choices of $\varepsilon = 3/(6 \cdot i + 5)$ with $i \in \mathbb{N}$, RANDOM needs at least $\Omega(n^\alpha \cdot \rho)$ queries to discover a ρ-fraction of the graph in expectation, where $\alpha > 0$ depends on ε.

3.1   The Case of Constant p

To prove that RANDOM discovers a ρ-fraction of the graph in expectation with only constantly many queries is straightforward. We start by showing a helpful lemma on queries and one node pair.

Lemma 1. *For a random graph $G_{n,p} = (V, E)$ and three distinct nodes $q, u, v \in V$, a query at q discovers the node pair u, v with probability at least $2 \cdot p \cdot (1 - p)$. The probability that k queries discover u, v is at least $x = 1 - (1 - 2 \cdot p \cdot (1 - p))^k$.*

Proof. We call a node $w \in V$ a *candidate*, if w is directly connected to v and not to u or directly connected to u and not to v. If the query node q is a candidate, it surely discovers the node pair u, v. This is independent of whether $\{u, v\} \in E$ or $\{u, v\} \in \overline{E}$. The probability of this event is $\Pr[q \text{ is candidate}] = 2 \cdot p \cdot (1 - p)$.

If we have several query nodes Q, the events "q is candidate" for $q \in Q$ are independent, since for each q the event depends on two distinct edges. Thus the probability that at least one query in Q discovers u, v is at least $\Pr[Q \text{ contains candidate}] = 1 - \Pr[\text{no } q \in Q \text{ is candidate}] = 1 - (1 - 2 \cdot p \cdot (1-p))^k$, where $k = |Q|$. □

The desired result is a corollary of the following theorem.

Theorem 1. *To discover a ρ-fraction of a $G_{n,p}$ graph in expectation, the RANDOM strategy needs at most $k = \lceil \log(1-\rho) / \log(1 - 2 \cdot p \cdot (1-p)) \rceil$ queries.*

Proof. By Lemma 1 we know that k queries Q discover a node pair $u, v \in V$ with probability at least $x = 1 - (1 - 2 \cdot p \cdot (1-p))^k$. The expected number of edges discovered by Q can be computed as $\mathbb{E}[\text{edges discovered by } Q] = \sum_{u,v \in V, u \neq v} \Pr[\{u, v\} \text{ is an edge and is discovered by } Q] \geq \sum_{u,v \in V, u \neq v} px = x \cdot \mathbb{E}[|E|]$. Similarly we obtain $\mathbb{E}[\text{non-edges discovered by } Q] \geq x \cdot \mathbb{E}[|\overline{E}|]$. Setting $x = \rho$ and solving for k gives the stated result. □

3.2 The Case of $p = n^\varepsilon / n$

Given an arbitrarily chosen constant $i \in \mathbb{N}$, in this entire section we set $\varepsilon = 3/(6 \cdot i + 5)$ and $p = n^\varepsilon / n$. By $\alpha, \beta, c > 0$ we always denote appropriately chosen constants, possibly depending on ε. Let $G_{n,p} = (V, E)$ be a random graph. By $U = \{u_1, \ldots, u_k\} \subset V$ we always denote an arbitrary node subset of constant cardinality. Let $N_i(U) := \bigcup_{\ell=1}^{k} N_i(u_\ell)$ denote the *i-neighborhood* of U. The *event A* plays a central role in our discussion and is defined as follows: for each $u \in U$ the size of its i-neighborhood is bounded from above by $|N_i(u)| \leq \bar{c} \cdot (np)^i$ and the size of its i-th layer is bounded from below by $|\Gamma_i(u)| \geq \underline{c}(np)^i$, for some constants $\bar{c}, \underline{c} > 0$. Additionally, there is no edge between the neighborhoods $N_i(u)$ and $N_i(v)$, for all pairs $u, v \in U$, $u \neq v$.

Lemma 2 states that event A holds w.h.p. Then in Lemma 3 we condition on event A and show that a node $w \in V \setminus N_i(U)$ is connected to two distinct i-th layers $\Gamma_i(u)$ and $\Gamma_i(v)$, for $u, v \in U$, with probability $\Omega(n^{-\beta})$, for a constant $\beta < 1$. We remark that the constant ε is chosen carefully on a "borderline": small enough such that Lemma 2 still holds and large enough for Lemma 3 to hold for some constant $\beta < 1$. These two lemmata can be applied to prove that a query node q is at the same distance $2 \cdot (i+1)$ from a node $u \in V$ and a node $v \in V$ w.h.p. Finally, we use this fact to show that $\Omega(n^\alpha \cdot \rho)$ queries are needed to discover a ρ-fraction of a $G_{n,p}$ graph in expectation, for some constant $\alpha > 0$. The proofs are based on two very helpful lemmata in [25] which give the tight bounds stated in event A on the size of the i-neighborhoods and the i-th layer.

Lemma 2. *Let i, ε, p be as given above. Let $G_{n,p} = (V, E)$ be a random graph and $U \subset V$ a constant cardinality node subset. Event A on $G_{n,p}$ and U holds with probability $1 - O(n^{-\alpha})$, for an appropriate constant $\alpha > 0$.*

Proof. We start by bounding the size of the neighborhoods and i-th layers. For a node $v \in V$ Lemma 2 from [25] states that $|N_i(v)| \leq \bar{c} \cdot (np)^i$ holds with probability at least $1 - o(n^{-1})$, for some constant $\bar{c} > 0$. To see that this bound actually holds simultaneously for all $u \in U$ with probability $1 - |U| \cdot o(n^{-1}) = 1 - o(n^{-1})$, simply consider the counter-events and apply the subadditivity of probabilities. Note that the cardinality $|U|$ is constant.

For a node $v \in V$ Lemma 8 from [25] states that if $G_{n,p}$ is connected, we have $|\Gamma_i(v)| \geq \underline{c}(np)^i$ with probability at least $1 - o(n^{-1})$, for some constant $\underline{c} > 0$. This bound actually holds simultaneously for all $u \in U$ with probability $1 - |U| \cdot o(n^{-1}) = 1 - o(n^{-1})$; again apply the subadditivity of probabilities to see this. Since $G_{n,p}$ is connected with probability at least $1 - o(n^{-1})$ for this range of p, cf. [26], the connectedness assumption can be dropped. In other words, the bounds on the size of the i-th layers of all $u \in U$ hold for any $G_{n,p}$ with probability $1 - o(n^{-1})$.

Combining both the bounds for the neighborhoods and the bounds for the i-th layers of the nodes in U, we have shown that the first part of event A holds with probability $1 - o(n^{-1})$.

We now come to the second part of event A. Let $x = \bar{c} \cdot (np)^i$ and consider c arbitrary node subsets of cardinality at most x. The probability that there is no edge from one of these subsets to another is at least

$$(1-p)^{\binom{c}{2} \cdot x^2} \geq (1-p)^{c^2 \cdot \bar{c}^2 (np)^{2i}} \geq \exp\left(-\frac{p}{1-p} \cdot c^2 \cdot \bar{c}^2 (np)^{2i}\right)$$

$$\geq \exp(-c' \cdot n^{\varepsilon-1} \cdot n^{2i\varepsilon})$$

for some constant $c' \geq c^2 \cdot \bar{c}^2/(1-p)$. With $\alpha = -(\varepsilon - 1 + 2i\varepsilon) = 2/(6 \cdot i + 5)$ we get

$$\exp(-c' \cdot n^{\varepsilon-1} \cdot n^{2i\varepsilon}) \geq 1 - c' \cdot n^{-\alpha}.$$

We conclude that with probability $1 - c' \cdot n^{-\alpha} = 1 - O(n^{-\alpha})$ there is no edge between any of the c subsets of cardinality x. To see that this also holds for the constant number $c = |U|$ of neighborhoods $N_i(u)$ as long as $|N_i(u)| \leq x$, for $u \in U$, we consider neighborhoods in iteratively defined subgraphs of the original $G_{n,p}$. Instead of $N_i(u_1)$ consider the neighborhood $N_i^{(1)}(u_1)$ in the random graph $G^{(1)} = G_{n,p} \setminus (U \setminus \{u_1\})$. For $1 < j \leq k$ we iteratively define the random graphs $G^{(j)} = G_{n,p} \setminus (\bigcup_{\ell \in \{1,\dots,j-1\}} N_i^{(\ell)}(u_\ell) \cup (U \setminus \{u_j\}))$ for which u_j's neighborhood is denoted by $N_i^{(j)}(u_j)$. By construction the neighborhoods $N_i^{(j)}(u_j)$ do not overlap and clearly $|N_i^{(j)}(u_j)| \leq |N_i(u_j)|$ holds, for $j \in \{1,\dots,k\}$. Moreover, by constructing $N_i^{(j)}(u_j)$ in such a way, no information about the edges between the individual $N_i^{(j)}(u_j)$ is revealed. Each such edge is still present with probability p. Therefore if $|N_i(u_j)| \leq x$, the computation goes through as above, yielding: with probability $1 - O(n^{-\alpha})$ there is no edge between any of the neighborhoods $N_i^{(j)}(u_j)$, $j \in \{1,\dots,k\}$. In this case we obviously have $N_i^{(j)}(u_j) = N_i(u_j)$.

Hence, the probability that the first and the second part of event A hold at the same time is at least $1 - o(n^{-1}) - O(n^{-\alpha}) = 1 - O(n^{-\alpha})$. Once more, this can be seen by considering the counter-events and applying the subadditivity of probabilities.   □

Lemma 3. *Let i, ε, p be as given above. Let $G_{n,p} = (V, E)$ be a random graph and $U \subset V$ a constant cardinality node subset. Conditioned on event A, a node $w \in V \setminus N_i(U)$ is connected to both $\Gamma_i(u)$ and $\Gamma_i(u')$ with probability $\Omega(n^{-\beta})$, for distinct $u, u' \in U$ and an appropriate constant $0 < \beta < 1$. This holds independently for all $w \in V \setminus N_i(U)$.*

Proof. Conditioning on event A reveals no information about the presence of edges between a node $w \in V \setminus N_i(U)$ and a node $v \in \Gamma_i(u)$, for $u \in U$. Such edges $\{w, v\}$ remain to be present independently with probability p.[1] Therefore, the probability that $w \in V \setminus N_i(U)$ is connected to both $\Gamma_i(u)$ and $\Gamma_i(u')$, for distinct $u, u' \in U$, is at least

$$(1 - (1-p)^{\underline{c}(np)^i})^2 \geq (1 - \exp(-p \cdot \underline{c}(np)^i))^2 \geq (1 - \exp(-\underline{c} \cdot n^{\varepsilon - 1 + i\varepsilon}))^2.$$

With $a = -(\varepsilon - 1 + i\varepsilon) = (3i + 2)/(6i + 5)$ this gives

$$(1 - \exp(-\underline{c} \cdot n^{-a}))^2 \geq \left(\frac{\underline{c} \cdot n^{-a}}{1 + \underline{c} \cdot n^{-a}} \right)^2 \geq \Omega(n^{-2a}) \geq \Omega(n^{-\beta}),$$

for some constant β, with $2a \leq \beta < 1$. Since the edges considered for different w do not overlap, this holds independently for all $w \in V \setminus N_i(U)$.   □

Lemma 4. *Let i, ε, p be as given above. Let $G_{n,p} = (V, E)$ be a random graph and $q, u, v \in V$ three distinct nodes. A query at q discovers the node pair u, v with probability $O(n^{-\alpha})$, for an appropriate constant $\alpha > 0$.*

Proof. For the graph $G_{n,p}$ and $U = \{u, v, q\}$ we assume that event A holds and under this assumption show that w.h.p. $d(q, u) = d(q, v) = 2(i + 1)$. In the following we concentrate on $d(q, u)$. Let $V_{q,u} \subseteq V \setminus N_i(\{u, v, q\})$ be some constant fraction of all nodes, i.e., $n_{q,u} = |V_{q,u}| \geq n/c'$ for some constant $c' > 1$. This is possible, since by event A we know $|N_i(\{u, v, q\})| = o(n)$.

To show that $d(q, u) = 2(i + 1)$ w.h.p., it suffices to show that at least one of the nodes in $V_{q,u}$ is connected to both $\Gamma_i(q)$ and $\Gamma_i(u)$ w.h.p. Note that by construction no node in $V_{q,u}$ can be connected with a node in $N_i(q) \setminus \Gamma_i(q)$ or $N_i(u) \setminus \Gamma_i(u)$. The probability that at least one node in $V_{q,u}$ is connected to both $\Gamma_i(q)$ and $\Gamma_i(u)$ by Lemma 3 and with appropriate constants $c, c'', \alpha' > 0, \beta < 1$ is at least

$$1 - (1 - c \cdot n^{-\beta})^{n_{q,u}} \geq 1 - \exp(-c \cdot n^{-\beta} \cdot n_{q,u})$$
$$\geq 1 - \exp(-c/c' \cdot n^{1-\beta}) \geq 1 - \exp(-c''n^{\alpha'}),$$

---

[1] Note on the other hand that by conditioning on event A we know that no edge between $w \in V \setminus N_i(U)$ and a node $v \in N_i(u) \setminus \Gamma_i(u)$, for $u \in U$, can be present. This will be used in the proof of Lemma 4.

With at least this probability $d(q, u) = 2(i + 1)$ holds. Note that by definition of $V_{q,u}$, there is still a constant fraction of all nodes left for $V_{q,v}$ and therefore $d(q, v) = 2(i + 1)$ holds w.h.p. as well.

Combining this with the probability for event A given by Lemma 2, we obtain that q is at the same distance from u and v with probability $1 - O(n^{-\alpha})$, for some constant $\alpha > 0$. Or equivalently: a query at q discovers the node pair u, v with probability at most $O(n^{-\alpha})$. □

We are now ready to state our main theorem.

Theorem 2. *Let $i \in \mathbb{N}$ be a given constant and set $\varepsilon = 3/(6 \cdot i + 5)$, $p = n^\varepsilon/n$. To discover a ρ-fraction of a $G_{n,p}$ graph in expectation, the RANDOM strategy needs at least $\Omega(n^\alpha \cdot \rho)$ queries, for some appropriately chosen constant $\alpha > 0$.*

Proof. Assume we need $k = o(n^\alpha \cdot \rho)$ random queries to discover a ρ-fraction in expectation. Let Q be a set of k queries returned by RANDOM. Then with Lemma 4 and $\mathbb{E}\left[|\overline{E}|\right] = \Omega(n^2)$ we get

$$\mathbb{E}\left[\text{non-edges discovered by } Q\right] \leq \sum_{q \in Q} \sum_{u,v \in V : u \neq v} \Pr\left[u, v \text{ discovered by } q\right]$$
$$\leq k \cdot O(n^{-\alpha}) \cdot n^2 = o(\rho) \cdot \mathbb{E}\left[|\overline{E}|\right].$$

This gives a contradiction and concludes the proof. □

3.3 Distances Within a Constant Cardinality Subset of the Nodes

We generalize the result in Lemma 4 to the case of distances between the nodes of an arbitrary constant cardinality subset of V: all distances are identical w.h.p. for certain choices of ε and $p = n^\varepsilon/n$. We believe this property is interesting in itself, since it does not necessarily seem intuitive at first sight.

Theorem 3. *With an arbitrarily chosen constant $i \in \mathbb{N}$, let $\varepsilon = 3/(6 \cdot i + 5)$ and $p = n^\varepsilon/n$. Let $G_{n,p} = (V, E)$ be a random graph and $U \subset V$ a constant cardinality node subset. All pairwise distances between the nodes in U are simultaneously equal to $2(i + 1)$ with probability $1 - O(n^{-\alpha})$, for an appropriate constant $\alpha > 0$.*

Proof. We proceed as in Lemma 4, but instead of just two, we define $\binom{|U|}{2}$ disjoint sets $V_{u,v} \subseteq V \setminus N_i(U)$ with $|V_{u,v}| \geq n/c'$, for $u, v \in U$ and some constant $c' > 1$. Note that such a constant exists, since $|U|$ is constant. The argumentation goes through for each $V_{u,v}$ independently as above, giving the statement of the theorem. □

4 Conclusion

We have introduced the notions of approximate network discovery and of discovering a large fraction of a graph in expectation. Motivated by previous computational experiments in random graphs, we have studied approximate network discovery in the $G_{n,p}$ model analytically for two different ranges of p. Surprisingly, we have been able to show that for constant p a constant number of queries suffices to discover a large fraction of a $G_{n,p}$ in expectation, whereas for certain small choices of p the number of queries chosen uniformly at random that discover a large fraction of $G_{n,p}$ grows with n. The analysis of the latter case also gave an interesting result for constant cardinality subsets of the nodes of a $G_{n,p}$, for certain small choices of p: w.h.p. all nodes of the subset are at exactly the same distance from each other.

A natural question for the case $p = n^\varepsilon/n$ is whether the lower bound that we have presented is tight. It would also be interesting to extend the analysis to other ranges of p. Analytically analyzing network discovery in scale-free random graphs would be of interest as well, in particular due to the practical relevance. Many real world networks (e.g., the Internet graph or peer-to-peer networks) are believed to have scale-free properties.

Acknowledgments. The authors would like to thank Martin Marciniszyn and Kostas Panagiotou for helpful discussions.

References

1. DIMES: Mapping the Internet with the help of a volunteer community (2004), http://www.netdimes.org
2. Oregon RouteViews: University of Oregon RouteViews project (1997), http://www.routeviews.org
3. Beerliová, Z., Eberhard, F., Erlebach, T., Hall, A., Hoffmann, M., Mihal'ák, M., Ram, L.S.: Network discovery and verification. In: Kratsch, D. (ed.) WG 2005. LNCS, vol. 3787, pp. 127–138. Springer, Heidelberg (2005)
4. Barrat, A., Hall, A., Mihal'ák, M.: Network discovery on snapshots of the Internet graph. Technical Report DELIS-TR-465, DELIS – Dynamically Evolving, Large-Scale Information Systems (2006)
5. Beerliová, Z., Eberhard, F., Erlebach, T., Hall, A., Hoffmann, M., Mihal'ák, M., Ram, L.S.: Network discovery and verification. IEEE Journal on Selected Areas in Communications 24(12), 2168–2181 (2006)
6. Barabási, A.L., Albert, R.: Emergence of scaling in random networks. Science 286, 509–512 (1999)
7. Cheswick, B., Burch, H.: Internet mapping project (1998), http://www.cs.bell-labs.com/who/ches/map/
8. Govindan, R., Reddy, A.: An analysis of Internet inter-domain topology and route stability. In: Proc. IEEE INFOCOM, April 1997, pp. 850–857. IEEE Computer Society Press, Los Alamitos (1997)
9. Govindan, R., Tangmunarunkit, H.: Heuristics for Internet map discovery. In: Proc. IEEE INFOCOM, March 2000, pp. 1371–1380. IEEE Computer Society Press, Los Alamitos (2000)
10. Gao, L.: On inferring autonomous system relationships in the Internet. IEEE/ACM Trans. Networking 9(6), 733–745 (2001)
11. Barford, P., Bestavros, A., Byers, J., Crovella, M.: On the marginal utility of deploying measurement infrastructure. In: Proc. ACM SIGCOMM Internet Measurement Workshop, November 2001, ACM Press, New York (2001)

12. Subramanian, L., Agarwal, S., Rexford, J., Katz, R.: Characterizing the Internet hierarchy from multiple vantage points. In: Proc. IEEE INFOCOM, IEEE Computer Society Press, Los Alamitos (2002)
13. Di Battista, G., Erlebach, T., Hall, A., Patrignani, M., Pizzonia, M., Schank, T.: Computing the types of the relationships between autonomous systems. IEEE/ACM Transactions on Networking 15(2), 267–280 (2007)
14. Achlioptas, D., Clauset, A., Kempe, D., Moore, C.: On the bias of traceroute sampling; or, power-law degree distributions in regular graphs. In: Proceedings of the 37th Annual ACM Symposium on Theory of Computing (STOC'05), pp. 694–703. ACM Press, New York (2005)
15. Dall'Asta, L., Alvarez-Hamelin, I., Barrat, A., Vázquez, A., Vespignani, A.: Statistical theory of Internet exploration. Phys. Rev. E 71 (2005)
16. Dall'Asta, L., Alvarez-Hamelin, I., Barrat, A., Vázquez, A., Vespignani, A.: Exploring networks with traceroute-like probes: theory and simulations. Theoret. Comput. Sci. 355(1), 6–24 (2006)
17. Khuller, S., Raghavachari, B., Rosenfeld, A.: Landmarks in graphs. Discrete Appl. Math. 70, 217–229 (1996)
18. Erlebach, T., Hall, A., Hoffmann, M., Mihal'ák, M.: Network discovery and verification with distance queries. In: Calamoneri, T., Finocchi, I., Italiano, G.F. (eds.) CIAC 2006. LNCS, vol. 3998, pp. 69–80. Springer, Heidelberg (2006)
19. Harary, F., Melter, R.A.: The metric dimension of a graph. Ars Combin., 191–195 (1976)
20. Chartrand, G., Zhang, P.: The theory and applications of resolvability in graphs: A survey. Congr. Numer. 160, 47–68 (2003)
21. Garey, M.R., Johnson, D.S.: Computers and Intractability: A Guide to the Theory of NP-Completeness. Freeman, San Francisco (1979)
22. Sebő, A., Tannier, E.: On metric generators of graphs. Math. Oper. Res. 29(2), 383–393 (2004)
23. Lindström, B.: On a combinatory detection problem I. Magyar Tud. Akad. Mat. Kutató Int. Közl. 9, 195–207 (1964)
24. Cáceres, J., Hernando, C., Mora, M., Pelayo, I.M., Puertas, M.L., Seara, C., Wood, D.R.: On the metric dimension of cartesian products of graphs. SIAM J. Discrete Math. 21(2), 423–441 (2007)
25. Chung, F., Lu, L.: The diameter of random sparse graphs. Adv. in Appl. Math. 26, 257–279 (2001)
26. Bollobás, B.: Random graphs. Academic Press, New York (1985)

A VNS Algorithm for Noisy Problems and Its Application to Project Portfolio Analysis

Walter J. Gutjahr, Stefan Katzensteiner, and Peter Reiter

Dept. of Statistics and Decision Support Systems, University of Vienna
{walter.gutjahr,stefan.katzensteiner,peter.reiter}@univie.ac.at
http://mailbox.univie.ac.at/walter.gutjahr/

Abstract. Motivated by an application in project portfolio analysis under uncertainty, we develop an algorithm S-VNS for solving stochastic combinatorial optimization (SCO) problems based on the Variable Neighborhood Search (VNS) metaheuristic, and show its theoretical soundness by a mathematical convergence result. S-VNS is the first general-purpose algorithm for SCO problems using VNS. It combines a classical VNS search strategy with a sampling approach with suitably increasing sample size. After the presentation of the algorithm, the considered application problem in project management, which combines a project portfolio decision on an upper level and project scheduling as well as staff assignment decisions on a lower level, is described. Uncertain work times require a treatment as an SCO problem. First experimental results on the application of S-VNS to this problem are outlined.

Keywords: Variable Neighborhood Search, stochastic combinatorial optimization, project portfolio selection, staff assignment, project scheduling.

1 Introduction

Stochastic combinatorial optimization (SCO) problems occur very frequently in diverse types of applications, since many optimization problems have a discrete or combinatorial structure, and very often, decisions have to be made under uncertainty. In this case, a standard approach offered by the operations research paradigm is to represent uncertain aspects quantitatively by a stochastic model, and to optimize the decision with respect to important characteristics (usually expected values, sometimes also variances, quantiles etc.) of the cost function which is then a random variable. For stochastic models involving a high degree of realism and therefore usually also a high degree of complexity, it can easily occur that these characteristics cannot be computed from explicit formulas or by means of numerical procedures, but must be estimated by simulation. This has led to the development of so-called *simulation-optimization* methods such as the Stochastic Branch-and-Bound Method [17], the Stochastic Ruler Method [2], or the Nested Partitions Method [18]. For a good survey on this area, cf. [5].

J. Hromkovič et al. (Eds.): SAGA 2007, LNCS 4665, pp. 93–104, 2007.
© Springer-Verlag Berlin Heidelberg 2007

Especially in cases where already the underlying deterministic combinatorial optimization (CO) problem is computationally hard and practical problem instances are of a medium or large size, combined methods treating the combinatorial aspect of the problem by a *metaheuristic* seem to be advisable. Such metaheuristic-supported simulation-optimization algorithms have been developed, e.g., based on Simulated Annealing ([7], [8], [1]), Genetic Algorithms [4], or Ant Colony Optimization ([9], [10]).

Recently, much attention is given in the literature to the *Variable Neighborhood Search* (VNS) metaheuristic developed by Hansen and Mladenovich [13] which seems to produce excellent results on several hard deterministic CO problems. To our best knowledge, however, no attempt has been made up to now to develop an extension of a VNS algorithm for the treatment of SCO problems by a simulation-optimization approach. Thus, a general-purpose technique for treating SCO problems by a VNS-based method seems still to be lacking.

In the present work, we propose a VNS-based algorithm called S-VNS which is applicable to a broad class of SCO problems. We give a formal "proof of concept" for the algorithm by deriving a strict mathematical *convergence result*: Under mild conditions, the current solutions of S-VNS converge to (global) optimality. A convergence result for a metaheuristic is certainly not yet a sufficient condition for its practical usefulness. However, one would hesitate to apply a heuristic algorithm for which cases cannot be excluded where the optimal solution can *never* be found, no matter how much computation time is invested. In a deterministic context, convergence of the "best solution so far" to the optimum can typically be easily achieved by introducing a sufficient amount of randomness into the search process. In a stochastic context, however, the problem is complicated by two factors: First, the "best solution so far" can not be determined with certainty by simulation. Secondly, the question of a suitable sample size scheme in an iterative stochastic optimization procedure is non-trivial, as outlined very clearly by Homem-de-Mello in [14].

The development of our VNS variant for stochastic problems is mainly motivated by an application in the field of project portfolio analysis under uncertainty. Practical project management, especially in the R & D area, requires (repeated) decisions on sets of projects, so-called *portfolios*, that are to be carried out, while rejecting or deferring other project candidates because of capacity limits or of insufficient benefits to be expected from those projects (cf. , e.g., [16] or [6]). In real applications, this decision cannot be separated from two other decisions to be made on a "lower decision level": one on the scheduling of the required work packages of each project that has been chosen, the other on the assignment of personnel to work packages. The problem is stochastic in so far as the work times needed by each work package have to be considered as random variables. Finally, when considering objective functions, we do not only take account of economic goals and of potential tardiness of projects, but also include a strategic goal connected with the work of the staff in different "competencies". All in all, this gives a problem formulation that is sufficiently complex to serve as a test case for the proposed S-VNS algorithm.

The plan of the paper is the following: In section 2, the proposed algorithm S-VNS is presented. Section 3 contains the mathematical convergence result. In section 4, we present the application problem under consideration, outline the way it is attacked, and present results of the application of S-VNS for a specific example instance. Section 5 contains conclusions.

2   The Proposed Simulation-Optimization Algorithm

2.1   General Problem Structure

The proposed algorithm is designed to provide solutions to the following rather general type of stochastic combinatorial optimization problems:

$$\text{Minimize } f(x) = \mathrm{E}\left(F(x, \omega)\right) \quad \text{s.t.} \quad x \in S.$$

Therein, x is the decision, F is the cost function, ω denotes the influence of randomness, E denotes the mathematical expectation, and S is a set of feasible decisions, which we always assume as finite.

We focus on application cases where it is difficult or even impossible to compute $\mathrm{E}\left(F(x, \omega)\right)$ numerically. Instead of determining approximations to this expected value by numerical techniques (e.g., numerical integration or analytical transform techniques), we resort to estimating it by sampling: For this purpose, a sample Z of s independent random *scenarios* $\omega_1, \ldots, \omega_s$ is drawn. The *sample average estimate* (SAE) is then given by $\tilde{f}(x) = (1/s) \sum_{\nu=1}^{s} F(x, \omega_\nu) \approx \mathrm{E}\left(F(x, \omega)\right)$. It is easy to see that $\tilde{f}(x)$ is an unbiased estimator for $f(x)$.

A typical application of our approach is given by the case where the SAEs are obtained by Monte-Carlo simulation. In this situation, we can interpret a scenario ω_ν as a random number or a vector of random numbers allowing, together with the solution x and the cost function F, the computation of an observation $F(x, \omega_\nu)$.

2.2   The Algorithm S-VNS

Our proposed solution algorithm, S-VNS, follows the general lines of the VNS algorithm as described in [13], but extends it by the computation of SAEs at several places, and by the introduction of an additional step called *tournament*, where a current candidate solution x is compared to a solution $\hat{x}$ considered best so far, replacing $\hat{x}$ by x if the SAE of x (based on an appropriate sample size) turns out as better than the SAE of $\hat{x}$.

As ordinary VNS, the procedure S-VNS uses a hierarchy of *neighborhood structures*: For each $x \in S$, we assume a *k-neighborhood* $\mathcal{N}_k(x) \subseteq S$ of x as given, where $k = 0, \ldots, k_{max}$. It is assumed that $\mathcal{N}_0(x) = \{x\}$ for all $x \in S$. Furthermore, we assume that the sets $\mathcal{N}_k$ $(k = 0, \ldots, k_{max})$ are nonempty and pairwise disjoint. The last assumption comes in a natural manner if the neighborhoods are derived from a distance function $d(.,.)$ on S by the definition $\mathcal{N}_k(x) = \{y \in S \,|\, d(x, y) = k\}$. (An alternative assumption would be to consider neighborhoods that form a chain of inclusions: $\mathcal{N}_0(x) \subset \mathcal{N}_1(x) \subset \ldots \subset \mathcal{N}_{k_{max}}$. This situation occurs if one sets $\mathcal{N}_k(x) = \{y \in S \mid d(x,y) \leq k\}$. Our theoretical result can also be extended to this case, but we skip this point for the sake of brevity.)

Procedure S-VNS

---

set $m = 1$;
choose $x \in S$;
set $\hat{x} = x$; // best-so-far solution
repeat until stopping criterion is met{
  (1) set $k = 1$;
  (2) **repeat** until $k = k_{max}$ {
    (a) (*shaking:*) generate an $x' \in \mathcal{N}_k(x)$ at random;
    (b) (*local search:*)
        set $\bar{x} = x'$;
        set local-optimum-found = false;
        set $\rho = 1$;
        repeat until local-optimum-found or $\rho > \rho_{max}$ {
            draw a sample Z of size s_0;
            compute the SAEs of $\bar{x}$ and of all solutions in $\mathcal{N}_1(\bar{x})$ w.r.t. Z;
            set $\bar{x}^* = $ solution in $\mathcal{N}_1(\bar{x})$ with best SAE;
            if ($\bar{x}^*$ has better SAE than $\bar{x}$) {
                set $\bar{x} = \bar{x}^*$;
                set $\rho = \rho + 1$;
            }
            else set local-optimum-found = true;
        }
        set $x'' = \bar{x}$;
    (c) (*move-or-not:*)
        draw a sample Z' of size s_0;
        compute the SAEs of x and of x'' w.r.t. Z';
        if (x'' has better SAE than x) {
            set $x = x''$;
            set $k = 1$;
        }
        else set $k = k + 1$;
    (d) (*tournament:*)
        draw a sample Z'' of size s_m;
        compute the SAEs of $\hat{x}$ and of x w.r.t. Z'';
        if (x has better SAE than $\hat{x}$)
            set $\hat{x} = x$;
        set $m = m + 1$;
  } }

Fig. 1. Pseudocode S-VNS

Fig. 1 shows the pseudo-code of S-VNS. We call the part (a) – (d) of the execution starting with a shaking step and ending with a tournament step a *round* of the algorithm. The integer m $(m = 1, 2, \ldots)$ is used as a round index. A round begins with the determination of a random neighbor solution x' to the current solution x ("shaking"), which aims at escaping from local optima. The size k of the neighborhood is dynamically controlled. The shaking mechanism must ensure that all solutions x' in the considered neighborhood have a positive probability of being selected. After that, x' is improved by local search steps until either a local optimum is found, or a certain predefined number ρ_{max} of steps is reached. Since the comparison of the current local search solution $\bar{x}$ with its neighbor solutions requires the estimation of objective function values, a sample Z (updated in each step) of constant size s_0 is used in each local search step.

Each round starts with some value of the neighborhood size variable k. We can group the rounds to *cycles*, where a new cycle is started any time when a round starts with $k = 1$. A cycle consists of one or several rounds with consecutively increasing values of k.

Local search results in a solution x''. In the step "move-or-not", this solution replaces the current incumbent x if it evaluates better at a further sample Z', again of size s_0. In the last case, the neighborhood size is reduced to one. Finally, the "tournament" between the incumbent x and the solution $\hat{x}$ considered best so far takes place. The size s_m of the sample Z'' on which this tournament is based has to be increased with suitable speed during the process in dependence on m. (This is contrary to the sizes of samples Z and Z' which are not increased.) We denote the solution x exposed to the tournament in round m by $x(m)$, and it will be shown in the next section that on certain conditions, the sequence $x(m)$ converges to a globally optimal solution.

3   Convergence Analysis

As it is seen from the pseudo-code, S-VNS works with different random samples at different occasions. Let us start their description with sample Z''. In round m, sample Z'' contains scenarios $\omega_1^m, \ldots, \omega_{s_m}^m$. We use the symbol ω^m for the s_m-tupel $(\omega_1^m, \ldots, \omega_{s_m}^m)$, and the symbol ω for $(\omega^1, \omega^2, \ldots)$.

Sample Z contains, in the ρ-th local search step of the m-th round, scenarios $\eta_{\rho,1}^m, \ldots, \eta_{\rho,s_0}^m$, and sample Z' contains scenarios which we denote by $\eta_{0,1}^m, \ldots, \eta_{0,s_0}^m$. By η^m, the matrix $(\eta_{\rho,\nu}^m)$ $(\rho = 0, \ldots, \rho_{max}; \nu = 1, \ldots, s_0)$ is denoted, and $\eta = (\eta^1, \eta^2, \ldots)$ comprises all η^m.

Finally, there is an additional influence of randomness which is active in the shaking step of the algorithm. We denote the random scenario influencing the choice of the solution x' in the shaking step of round m by ξ^m, and set $\xi = (\xi^1, \xi^2, \ldots)$. Note that apart from ξ^m, the decision on x' is also dependent on the current x and on the current value of k.

The total information (ξ, η, ω) determining the random influence will be called *sample path*.

Throughout this article, we assume that all scenarios ω_ν^m, $\eta_{\rho,\nu}^m$ and ξ^m are independent. Moreover, it is assumed that the ω_ν^m have identical probability measures for all m and ν, the $\eta_{\rho,\nu}^m$ have identical probability measures for all m, ρ and ν, and the ξ^m have identical probability measures for all m. These assumptions are satisfied in a natural way if S-VNS applies Monte-Carlo simulation for cost evaluation, based on a series of independent random numbers the distribution of which is independent of x and m.

Note that we have two different probability mechanisms in the algorithm. The first mechanism generates the part (ξ, η) of the sample path. This mechanism completely determines the control flow in S-VNS, with the exception of the question whether in the tournament, $\hat{x}$ is updated to the value x or not. We denote the probability space consisting of the elements (ξ, η) by Ξ, and the probability measure on it by P. The second mechanism generates the part ω of the sample path. It influences the decision which solution wins a tournament. We denote the probability space consisting of the elements ω by Ω, and the probability measure on it by $\tilde{P}$.

For deriving our main theoretical result, we make use of the following lemma which has be proven by large-deviation theory:

Lemma 3.1. (Homem-de-Mello [14], Proposition 3.2). Suppose that for a scheme $(s_1, s_2, \ldots)$ of sample sizes and independent scenarios ω_ν^m,

(i) for each $x \in S$, the variances $var[F(x, \omega_1^m)]$ are bounded by some constant $M = M(x) > 0$,

(ii) the scenarios ω_ν^m have identical probability measures, and the SAEs $\tilde{f}_m(x) = (1/s_m) \sum_{\nu=1}^{s_m} F(x, \omega_\nu^m)$ are unbiased[1], i.e., $E(\tilde{f}_m(x)) = f(x)$ for all x,

(iii) $\sum_{m=1}^\infty \alpha^{s_m} < \infty$ for all $\alpha \in]0, 1[$.

Then for each x, we have $\tilde{f}_m(x) \to f(x)$ $(m \to \infty)$ for $\tilde{P}$-almost all $\omega \in \Omega$.

We are now able to show convergence of S-VNS to a (globally) optimal solution. For the proof, we refer to the technical report [12].

Theorem 3.1. Consider S-VNS. Suppose that

(a) $\bigcup_{k=0}^{k_{max}} \mathcal{N}_k(x) = S$ for all $x \in S$,

(b) the assumptions (i) – (iii) of Lemma 3.1 are satisfied for the tournaments in S-VNS, and

(c) the distribution of $F(x, \omega)$ has a nonzero density in each point of $\mathbb{R}$.

Then, for P-almost all $(\xi, \eta) \in \Xi$ and $\tilde{P}$-almost all $\omega \in \Omega$, there exists an integer m_0 (depending on the sample path (ξ, η, ω)) such that $x_m \in S^*$ for all $m \geq m_0$, where S^* is the set of optimal solutions.

Remark 1. In typical VNS applications, k_{max} is usually set to a smaller value than the minimum value ensuring that each solution can be reached from any x

---

[1] Homem-de-Mello also refers to a more general framework where $E(F(x, \omega_\nu^m))$ can be different from $f(x)$. In our context, unbiasedness follows by definition.

within a k-neighborhood with $k \leq k_{max}$, as assumed in condition (a) of Theorem 3.1. In future investigations, one might try to generalize the result by omitting condition (a).

Remark 2. A sufficient condition for assumption (iii) in Lemma 2.1 to be satisfied is that the sample size s_m grows as $c_0 \cdot \sqrt{m}$ with a constant $c_0 > 0$, which is a fairly moderate growth. However, not all increasing samples size schedules satisfy assumption (iii); e.g., logarithmic growth is not sufficient.

4   The Application Problem: Stochastic Project Selection, Scheduling and Staffing

4.1   Problem Formulation

We apply the proposed algorithm to the stochastic version of a problem the deterministic version of which has been introduced in [11] (cf. also [16], [6]). The problem encompasses a project portfolio selection decision on an upper decision level and decisions on project scheduling and staff assignment on a lower level. The upper-level decision consists in the choice of a subset of projects from a given set of candidates $i = 1, \ldots, n$. Let the binary decision variable x_i take the value 1 iff project i is to be selected, and let $x = (x_1, \ldots, x_n)$. Projects have ready times ρ_i and due dates δ_i. The decision has to be made under uncertainty on the work times needed by the projects with respect to certain required human competencies $r = 1, \ldots, R$, which can be considered as "resources". It is assumed that project i requires a (so-called "effective") work time of D_{ir} in competency r $(r = 1, \ldots, R)$, where D_{ir} is a random variable with a (known) distribution the parameters of which can be estimated in advance. We call that part of a project i that requires a particular competency r the *work package* with index (i, r).

A fixed team of employees $j = 1, \ldots, J$ is assumed as available. For each employee, we suppose that her/his efficiency in each competency r can be quantified as a value γ_{jr}, measuring the fraction of effective work in competency r that s/he is able to deliver within given time, compared to the work of an employee with a "perfect" ability in the considered competency r. If employee j with efficiency γ_{jr} works for y time units in work package (i, r), s/he reduces the effective work time required for work package (i, r) by the amount $\gamma_{jr} \cdot y$. To distinguish y from the effective work time $\gamma_{jr} \cdot y$, we call y the *real* work time.

On the lower decision level, after a decision on a project portfolio has been made, the workload corresponding to the single competencies has to be assigned to the staff over time. Since work times are uncertain, a dynamic policy is required for this part of the planning process. We will consider a specific, fixed policy of this type. It is allowed that several employees contribute to the same work package (i, r), provided that the efficiency value of each of these employees in competency r does not lie below some pre-defined minimum value γ_{min}. Each contribution is weighted by the efficiency value of the assigned employee.

As the objective function, we take a weighted average with fixed weights β_1, β_2 resp. β_3 of three terms: (1) Economic benefit (return etc.), which can be estimated for each project i by a value w_i, its overall value resulting as the sum of the values w_i of the selected projects i. (2) Expected strategic benefit, where strategic benefit measures the degree to which the company engages in future-oriented areas. To quantify this objective, desirability values v_r are assigned to each competency r. The strategic benefit is defined as the weighted sum of the real work times $Y_r(x)$ the overall staff spends in each competency r (given portfolio x), weighted by the values v_r. It is important to note that we take here real work times instead of effective work times; the reason for this choice is that in models on organizational learning (cf., e.g., [3]), competence increment is rather related to real than to effective work. As a consequence, the second objective function is not simply determined by the set of selected projects, but influenced by the staff assignment decision. Observe that the quantities $Y_r(x)$ depend also on the random variables D_{ir} and are therefore random variables themselves. (3) The expected value of the total tardiness $\Psi(x)$ of the selected projects with respect to their due date (given portfolio x). The value $\Psi(x)$ is the sum of the tardiness values $\Psi_i(x)$ of the selected projects i, where, with $C_i(x)$ denoting the completion time of project i under portfolio decision x, the tardiness $\Psi_i(x)$ is defined as $(C_i(x) - \delta_i)^+$ if $x_i = 1$, and zero otherwise. Objective (3) enters with negative sign into the overall objective function, since it is to be minimized.

In total, this produces the objective function

$$\mathrm{E}\left(\beta_1 \sum_{i=1}^{n} w_i x_i + \beta_2 \sum_{r=1}^{R} v_r Y_r(x) - \beta_3 \sum_{i=1}^{n} \Psi_i(x)\right) \to \max,$$

where the first term is in fact deterministic. The constraint is defined by the (given) capacity limits of the employees.

4.2   Solution Technique

On the upper decision level of project portfolio selection, the problem presented in the previous subsection is an SCO problem, which makes the proposed algorithm S-VNS applicable. The feasible set S is in this case the set $\{0,1\}^n$ of possible project portfolios. What remains to be specified is the way the two other aspects of the problem, namely project scheduling and staff assignment, are handled on the lower decision level.

Evidently, our problem formulation contains a (particular) *stochastic scheduling problem* as a special case. Methods for solving stochastic scheduling problems have found considerable interest in the literature; we refer, e.g., to Möhring and Stork [15]. As outlined in [15], a suitable approach to the solution of a stochastic scheduling problem consists in determining a *policy* which acts at so-called *decision points* of the time axis. Roughly speaking, a decision point is a time point where "something happens". In our case, decision points are $t = 0$ (project start), the ready times ρ_i of the projects, and the (random) completion times of the work packages (i, r). It is important to ensure that the decision prescribed by some policy for decision point t only uses information that is already available at time t; for example, the knowledge of the random work time D_{ir} required by a work package (i, r) that is not yet completed at time t cannot be used in the decision. However, since we assume that the *distributions* of these random variables are known in advance, quantities as $E(D_{ir})$ may be used by the policy at any time.

Procedure Stochastic Scheduling-and-Staffing

---

set $\tau = 0$; // decision point
repeat until all work packages are completed {
    for all projects i in ascending order of their due dates δ_i {
        for all yet uncompleted work packages (i, r) with $\tau \geq \rho_i$ in descending order
        of $E(D_{ir})$ {
            for all employees j in descending order of γ_{jr} {
            if (employees j is not yet assigned and $\gamma_{jr} \geq \gamma_{min}$)
                assign employee j to work package (i, r);
    } } }
    determine the earliest time τ' at which either one of the currently scheduled
      work packages becomes completed, or a ready time occurs;
    subtract, for each work package, the effective work done between time τ and
      time τ' from the remaining effective work time (work times by employees
      have to be weighted by their efficiencies);
    set $\tau = \tau'$, and set all assignments of employees to work packages back;
}

Fig. 2. Dynamic policy for the lower decision level

For scheduling the remaining work in the selected projects and to re-assign staff in a decision point, we have chosen a conceptually simple heuristic policy which can be classified as *priority-based* in the terminology of [15]: First of all, projects with an earlier due date have a higher priority of being scheduled immediately than projects with a later due date. Secondly, within a project i, work packages (i, r) with a higher value of $E(D_{ir})$ (i.e., those that can be expected to consume a higher amount of resources) obtain higher priorities. Finally, for each work package (i, r), employees j with free capacities *and* efficiency value $\gamma_{jr} \geq \gamma_{min}$ are assigned according to priorities defined by their efficiency value γ_{jr} (higher priority for higher efficiency). We present the applied stochastic scheduling-and-staffing algorithm in pseudo-code form in Fig. 2. The chosen policy effects that at each decision point t, the set of employees is partitioned into teams where each team is either assigned to a specific work package, or is (in the case at most one team) "idle" at the moment.

4.3   A Numerical Example

At the moment, we are performing tests using benchmark data on candidate projects, employees, work times, efficiencies and objectives from an application provided by the E-Commerce Competence Center Austria ("EC3"), a public-private R & D organization. The following illustration example is a simplified version derived from this real-world application. Experiments were performed on a PC Pentium 2.4 GHz with program code implemented in MATLAB V 6.1.

The reduced example application (which does not use the EC3 data directly, but is similar in flavor) considers $n = 12$ projects, $J = 5$ employees and $R = 3$ competencies. Ready times and due dates span over a time horizon of 32 months. For modelling the work times D_{ir}, triangular distributions have been used for the first tests, where the parameter estimations map the typical high skewness of work times in R & D projects, the most frequent value lying much closer to the minimum value than to the maximum value. (We do not claim that triangular distribution satisfy the conditions of Theorem 3.1, but they yield meaningful test cases for the algorithm S-VNS nevertheless.) Efficiency values have been determined by a special procedure the description of which is outside the scope of this paper; in the case of the five selected employees for the illustration example presented here, the efficiency values are such that one of the five employees is highly specialized, whereas the efficiencies of the other four employees are more evenly distributed over the three considered competencies. A minimum efficiency score of $\gamma_{min} = 0.25$ has been defined. The economic benefits of the 12 projects have roughly been estimated by the vector $(20, 1, 4, 5, 4, 4, 4, 4, 3, 1, 6, 8)$; for the strategic benefits, the three considered competencies have been weighted by the desirability values 2, 1 and 5, respectively. The time unit is a month. First, we considered the case where the weights for the three parts of the objective function (corresponding to economic benefit, strategic benefit and tardiness) are estimated by the numbers $\beta_1 = 20$, $\beta_2 = 1$ and $\beta_3 = 100$, respectively. This means that work experience of 20 months gained in a competency with desirability 1 is considered as of the same worth as an additional unit of economic benefit, and the loss caused by a delay of one month is considered as equivalent to the loss of five units of economic benefit.

It is not easy to evaluate the results delivered by the proposed algorithm S-VNS, since exact solutions[2] are very hard to determine already for this comparably small problem size: Note that the solutions space for the portfolio decision consists of $2^{12} = 4096$ portfolios, and from the observed variances and the typical differences between objective function values, we estimate that at least 10^4 simulation runs are necessary to compute a sufficiently accurate approximation to the true objective function value $f(x)$ enabling an identification of the optimal solution with some reliability. Of course, even this time-consuming brute-force procedure would not give a guarantee that the true optimum is found. In order to have a yardstick for our heuristic results, we adopted the following pragmatic procedure instead: First, a heuristic solution $\tilde{x}$ was determined by S-VNS, and using the observed SAE variance, an aspiration level c_a was determined in such way that an SAE with sample size $s^{(1)}$ would be above c_a for each solution that is at least as good as $\tilde{x}$, except with a very small failure probability $< 10^{-6}$. Then, we performed complete enumeration of all portfolios $x \in S$ by an SAE with

---

[2] By the term "exact solution", we refer in the sequel only to the portfolio selection decision, considering the described scheduling-and-staffing policy as fixed. The question whether this policy can be improved will not be treated here.

sample size $s^{(1)}$. Whenever the resulting objective function estimate exceeded the aspiration level c_a for a solution x, we re-evaluated this solution based on a second SAE with large sample size $s^{(2)} \gg s^{(1)}$. For $s^{(1)}$ and $s^{(2)}$, we chose the values 80 and 20 0000, respectively.

By this technique, we found the presumably best solution within a runtime of about 5 hours and 20 minutes. For the weights indicated above, this was the solution $x^* = (0, 0, 1, 1, 1, 1, 0, 1, 1, 1, 0, 1)$. We performed now runs of S-VNS where the stopping criterion was a pre-defined maximum number m_{max} of rounds chosen in such a way that each run required about 1 minute computation time only. For each parameter combination, the average relative deviation of the objective function values of the delivered solutions in ten independent runs from that of the (presumably) best solution was computed. We observed that for several parameter combinations, the optimal solution was found in a majority of runs. Let us look at two special parameter combinations. Therein, c_0 denotes the factor of $\sqrt{m}$ in Remark 2 after Theorem 3.1, and Δ_r gives the average relative deviation of the solution quality from the optimum in percent. It can be seen that the relative derivations are rather low:

(i) $s_0 = 5$, $c_0 = 5$, $k_{max} = 12$, $\rho_{max} = 10$, $m_{max} = 10$: $\Delta_r = 1.13\%$.
(ii) $s_0 = 3$, $c_0 = 10$, $k_{max} = 12$, $\rho_{max} = 10$, $m_{max} = 10$: $\Delta_r = 0.86\%$.

Reducing the weight β_3 for objective function 3 from 100 to 20 describes the situation where violations of due dates have less serious consequences. The optimal solution for this weight combination is $x^* = (1, 0, 1, 1, 1, 1, 1, 1, 1, 0, 0, 0)$. It is seen that here, the decision maker can risk to include the larger (and more profitable) project no. 1 into the portfolio. Using parameter combination (i) from above, all of ten independent runs of S-VNS provided this optimal solution.

5   Conclusions

We have presented a simulation-optimization algorithm S-VNS for stochastic combinatorial optimization problems, based on the Variable Neighborhood Search metaheuristic, and shown that the solutions produced by S-VNS converge under rather weak conditions to optimality. As an application case, a stochastic project portfolio selection problem involving project scheduling and staff assignment decisions has been described, and the application of the proposed algorithm to this problem has been outlined.

First experimental observations have been obtained, which are of course not yet sufficient for an experimental evaluation of the approach, but show promising results. Tests on a large number of randomly generated instances as well as on real-life instances will be necessary to judge the solution quality achieved by the method. In particular, these tests should include comparisons with other simulation-optimization techniques and verify the found results by statistical significance tests. Within our mentioned application context, we are performing extensive investigations of this type, and a long version of the present paper is planned providing results of this type.

Acknowledgment. Financial support from the Austrian Science Fund (FWF) by grant #L264-N13 is gratefully acknowledged.

References

1. Alrefaei, M.H., Andradottir, S.: A Simulated Annealing algorithm with constant temperature for discrete stochastic optimization. Management Sci. 45, 748–764 (1999)
2. Alrefaei, M.H., Andradottir, S.: A modification of the stochastic ruler method for discrete stochastic optimization. European J. of Operational Research 133, 160–182 (2001)
3. Chen, A.N.K., Edgington, T.M.: Assessing value in organizational knowledge creation: considerations for knowledge workers. MIS Quaterly 29, 279–309 (2005)
4. Fitzpatrick, J.M., Grefenstette, J.J.: Genetic algorithms in noisy environments. Machine Learning 3, 101–120 (1988)
5. Fu, M.C.: Optimization for simulation: theory vs. practice. INFORMS J. on Computing 14, 192–215 (2002)
6. Gabriel, S.A., Kumar, S., Ordonez, J., Nasserian, A.: A multiobjective optimization model for project selection with probabilistic considerations. Socio-Economic Planning Sciences 40, 297–313 (2006)
7. Gelfand, S.B., Mitter, S.K.: Simulated Annealing with noisy or imprecise measurements. J. Optim. Theory Appl. 69, 49–62 (1989)
8. Gutjahr, W.J., Pflug, G.: Simulated annealing for noisy cost functions. J. of Global Optimization 8, 1–13 (1996)
9. Gutjahr, W.J.: A converging ACO algorithm for stochastic combinatorial optimization. In: Albrecht, A.A., Steinhöfel, K. (eds.) SAGA 2003. LNCS, vol. 2827, pp. 10–25. Springer, Heidelberg (2003)
10. Gutjahr, W.J.: S-ACO: An ant-based approach to combinatorial optimization under uncertainty. In: Dorigo, M., Birattari, M., Blum, C., Gambardella, L.M., Mondada, F., Stützle, T. (eds.) ANTS 2004. LNCS, vol. 3172, pp. 238–249. Springer, Heidelberg (2004)
11. Gutjahr, W.J., Katzensteiner, S., Reiter, P.: Stummer, Ch., Denk, M., Competence-driven project portfolio selection, scheduling and staff assignment, Tech. Rep. Univ. of Vienna, Dept. of Statistics and Decision Support Systems (2007)
12. Gutjahr, W.J., Katzensteiner, S., Reiter, P.: Variable neighborhood search for noisy problems applied to project portfolio analysis, Tech. Rep., Univ. of Vienna, Dept. of Statistics and Decision Support Systems (2007)
13. Hansen, P., Mladenović, N.: Variable neighborhood search: Principles and applications. European J. of Operational Research 130, 449–467 (2001)
14. Homem-de-Mello, T.: Variable-sample methods for stochastic optimization, ACM Trans. on Modeling and Computer Simulation 13, 108–133 (2003)
15. Möhring, R.H., Stork, F.: Linear preselective policies for stochastic project scheduling. Mathematical Methods of Operations Research 52, 501–515 (2000)
16. Nozic, L.K., Turnquist, M.A., Xu, N.: Managing portfolios of projects under uncertainty. Annals of Operations Research 132, 243–256 (2004)
17. Norkin, V.I., Ermoliev, Y.M., Ruszczynski, A.: On optimal allocation of indivisibles under uncertain. Operations Research 46, 381–395 (1998)
18. Shi, L., Olafsson, S.: Nested partitions method for global optimization, Operations Research 48, pp. 390–407 (2000)

Digit Set Randomization in Elliptic Curve Cryptography

David Jao[1], S. Ramesh Raju[2,3], and Ramarathnam Venkatesan[3,4]

[1] University of Waterloo, Waterloo ON N2L3G1, Canada
djao@math.uwaterloo.ca
[2] Theoretical Computer Science Lab, IIT Madras, Chennai – 600036, India
srraju@cse.iitm.ernet.in
[3] Microsoft Research India Private Limited, "Scientia", No:196/36,
2nd Main Road, Sadashivnagar, Bangalore – 560080, India
[4] Microsoft Research, 1 Microsoft Way, Redmond WA 98052, USA
venkie@microsoft.com

Abstract. We introduce a new approach for randomizing the digit sets of binary integer representations used in elliptic curve cryptography, and present a formal analysis of the sparsity of such representations. The motivation is to improve the sparseness of integer representations and to provide a tool for defense against side channel attacks. Existing alternative digit sets D such as $D = \{0, 1, -1\}$ require a certain non-adjacency property (no two successive digits are non-zero) in order to attain the desired level of sparseness. Our digit sets do not rely on the non-adjacency property, which in any case is only possible for a certain very restricted class of digit sets, but nevertheless achieve better sparsity. For example, we construct a large explicit family of digit sets for which the resulting integer representations consist on average of 74% zeros, which is an improvement over the 67% sparsity available using non-adjacent form representations. Our proof of the sparsity result is novel and is dramatically simpler than the existing analyses of non-adjacent form representations available in the literature, in addition to being more general. We conclude with some performance comparisons and an analysis of the resilience of our implementation against side channel attacks under an attack model called the *open representation model*. We emphasize that our side channel analysis remains preliminary and that our attack model represents only a first step in devising a formal framework for assessing the security of randomized representations as a side channel attack countermeasure.

Keywords: randomized representations, elliptic curve cryptography, non-adjacent form representations, side channel attack countermeasures.

1 Introduction

Let α be an elliptic curve private key. In traditional elliptic curve cryptography, a point of the form αQ is computed via repeated doubling and addition using the binary representation $\alpha = a_k a_{k-1} \ldots a_1 a_0$ of α. By exploiting the fact that

J. Hromkovič et al. (Eds.): SAGA 2007, LNCS 4665, pp. 105–117, 2007.
© Springer-Verlag Berlin Heidelberg 2007

inverses on an elliptic curve are easy to compute, one can speed up the computation of αQ using signed binary representations [2,?]. As a simple example, consider the case where the integers a_i are taken from the set $\{0, 1, -1\}$. In this case, the resulting representations $\alpha = \sum_{i=0}^{k} a_i 2^i$ are no longer unique, but Reitweisner [22] observed in 1960 that these representations become unique if one decrees that no two consecutive a_i are nonzero. The resulting representations are known as *non-adjacent form* representations or NAF representations in the literature. Furthermore, the NAF representation of α is guaranteed to have the fewest possible nonzero terms out of all possible representations of α using $\{0, 1, -1\}$, a property which is desirable for performance reasons because nonzero terms slow down the computation of αQ. Morain and Olivos [14] were among the first to exploit $\{0, 1, -1\}$-representations to speed up elliptic curve computations.

Recently, Muir and Stinson [15] studied representations of the form $\alpha = \sum_{i=0}^{k} a_i$, where $a_i \in \{0, 1, x\}$ for some constant x, and found an infinite (but exponentially rare) class of sets $\{0, 1, x\}$, called *non-adjacent digit sets* or NADS, satisfying the property that each integer α has a unique NAF representation in $\{0, 1, x\}$. Subsequent work [1,7] has extended the understanding of the properties of NADS and their corresponding NAF representations, but such research has had at best limited applicability to cryptography because of the exponential rarity of known NADS, which results in only a limited variety of such sets being available for use in implementations.

In this paper we introduce and study binary representations with respect to more general digit sets of the form $\{0, 1, x, y, \ldots z\}$. We show that the high performance characteristics of traditional signed binary representations can be realized over this much larger and more general collection of digit sets. Our result enables an entire new class of algorithms for runtime randomization of elliptic curve exponentiation, based on randomized digit sets. We provide both theoretical and empirical analysis showing that EC exponentiation using randomized sparse representations is superior to traditional exponentiation or signed exponentiation in efficiency. Our theoretical analysis is simpler than prior investigations even when restricted to the special case of non-adjacent digit sets of the form $\{0, 1, x\}$, but our results also apply more generally to digit sets having size $2^c + 1$ for any c, with only mild restrictions (e.g. the set must contain one element congruent to 3 mod 4). We achieve this ease of analysis by allowing the use of integer representations which occasionally violate the nonadjacency rule. Nevertheless, we show that these representations have zero density asymptotically equal to or better than the uniquely defined representations arising from NADS. Finally, we provide an analysis indicating that the information entropy of an integer multiplier is lower bounded by that of the digit set under an attack model which we call the *open representation model*, in which the symbolic representation of the integer multiplier (that is, the pattern of digits appearing in the representation) is exposed to the attacker via side channel information but the digit set itself is hidden. The use of randomized digit sets is crucial to this analysis, because otherwise there is no distinction between knowing the symbolic representation of an integer and knowing the integer itself. Since all known side channel attacks to date (for example, [11,?,?]) operate by obtaining the symbolic representation of an integer, we believe that the introduction of randomized digit sets and the creation of such a distinction under the open representation model constitutes a crucial first step in devising a rigorous framework for analyzing side channel attack countermeasures. We emphasize, however, that our preliminary investigations fall short of a complete framework for side channel attack analysis and that much more remains to be done in this area.

2   Statistical Properties of NAF Representations

Heuberger and Prodinger [7] recently showed that non-adjacent form representations with respect to digit sets $\{0, 1, x\}$ have an average density of nonzero terms equal to $1/3$, using a detailed combinatorial study involving recurrences. In this section we give a Markov Chain analysis for the $\{0, 1, x\}$ case, which as we will see generalizes readily to larger digit sets. We begin with the relevant definitions.

Definition 2.1. *A* digit set *is a finite set of integers containing both* 0 *and* 1 *as elements.*

For the rest of this section, we assume the digit set D has the form $D = \{0, 1, x\}$ where $x \equiv 3 \pmod 4$ is negative.

Definition 2.2. *Let D be a digit set and let α be a nonnegative integer. A non-adjacent form* representation *(or* NAF *representation) of α with respect to D is a finite (possibly empty) sequence of integers $a_i \in D, i = 0, \ldots, k$, with $a_k \neq 0$, such that $\alpha = \sum_{i=0}^{k} a_i 2^i$, and no two consecutive values of a_i are both nonzero.*

We note at this point that an arbitrary integer α does not necessarily have a NAF representation with respect to D. For the moment, we will limit our attention to the case where α does have a NAF representation with respect to D. Later we will discuss how to modify our algorithm and analysis to apply to the cases where it does not.

Theorem 2.3 ([15]) *Every nonnegative integer has at most one* NAF *representation with respect to D.*

We write $\alpha = (a_k \cdots a_2 a_1 a_0)_2$ to denote that the sequence a_i is the NAF representation for α. By convention, the empty sequence is the NAF representation for 0.

The following definition and theorem provide a method for computing NAF representations.

Definition 2.4. *For any digit set $D = \{0, 1, x\}$, let $f_D \colon \mathbb{N} \to \mathbb{N}$ and $g_D \colon \mathbb{N} \to D \cup (D \times D)$ be the functions defined by*

$$
f_D(n) = \begin{cases} n/4 & n \equiv 0 \pmod 4 \\ (n-1)/4 & n \equiv 1 \pmod 4 \\ n/2 & n \equiv 2 \pmod 4 \\ (n-x)/4 & n \equiv 3 \pmod 4 \end{cases}, \qquad g_D(n) = \begin{cases} (0,0) & n \equiv 0 \pmod 4 \\ (0,1) & n \equiv 1 \pmod 4 \\ 0 & n \equiv 2 \pmod 4 \\ (0,x) & n \equiv 3 \pmod 4 \end{cases}.
$$

Theorem 2.5 ([15]) *A nonnegative integer α has a* NAF *representation if and only if $f_D(\alpha)$ has a* NAF *representation. Moreover, if $\alpha = (a_k \cdots a_2 a_1 a_0)_2$ and $f_D(\alpha) = (b_\ell \cdots b_2 b_1 b_0)_2$, then $a_k \cdots a_2 a_1 a_0 = b_\ell \cdots b_2 b_1 b_0 \parallel g_D(\alpha)$, where $\parallel$ denotes concatenation of sequences.*

Theorem 2.5 suggests the following algorithm $\mathcal{A}$ for computing the NAF representation of α:

1. If $\alpha = 0$, then return the empty string. Otherwise:
2. Evaluate $f_D(\alpha)$ and $g_D(\alpha)$.
3. Recursively call the algorithm $\mathcal{A}$ on the new input value $f_D(\alpha)$ in order to find the NAF representation of $f_D(\alpha)$.
4. Concatenate the NAF representation of $f_D(\alpha)$ with $g_D(\alpha)$, and remove any leading zeros, in order to obtain the NAF representation for α.

By Theorems 2.3 and 2.5, the algorithm $\mathcal{A}$ is guaranteed to return the NAF representation of α whenever α has one.

2.1 $\mathcal{A}$ As a Dynamical System

The execution profile of the algorithm $\mathcal{A}$ involves calculating the quantities $\alpha_1 = f_D(\alpha)$, $\alpha_2 = f_D(\alpha_1) = f_D^2(\alpha)$, $\alpha_3 = f_D(\alpha_2) = f_D^3(\alpha)$, etc., as well as the values of $g_D(\alpha)$, $g_D(\alpha_1)$, $g_D(\alpha_2)$, etc. We are interested in knowing the distribution of the integers $\alpha_k \bmod 4$ in order to predict which of the execution pathways for f_D and g_D in Definition 2.4 are more likely to be encountered.

Theorem 2.6 *For a fixed digit set $D = \{0, 1, x\}$, where $x \equiv 3 \pmod 4$ the probability distribution of the congruence classes $\alpha_k \bmod 4$ over the values $(0, 1, 2, 3)$, for random uniformly selected integers $\alpha \in [0, N]$, where $N \gg |x|$, converges to the vector $(\frac{1}{5}, \frac{3}{10}, \frac{1}{5}, \frac{3}{10})$ as $k \to \infty$, with error bounded in magnitude by an exponential in k.*

Proof. By hypothesis, the initial (uniformly selected) input value α has probability distribution $\mathcal{P}_0 = (\frac{1}{4}, \frac{1}{4}, \frac{1}{4}, \frac{1}{4})$ over the congruence classes mod 4. The probability distribution $\mathcal{P}_1$ for α_1 is computed as follows:

- By assumption, α is uniformly distributed in $[0, N]$.
- If $\alpha \equiv 0 \pmod 4$, then $\alpha_1 = \alpha/4$ is uniformly distributed mod 4.
- If $\alpha \equiv 1 \pmod 4$, then $\alpha_1 = \frac{\alpha-1}{4}$ is uniformly distributed mod 4.
- If $\alpha \equiv 2 \pmod 4$, then $\alpha_1 = \alpha/2$ is uniformly either 1 or 3 mod 4.
- If $\alpha \equiv 3 \pmod 4$, then $\alpha_1 = \frac{\alpha-x}{4}$ is uniformly distributed mod 4.

Denote by A the matrix

$$
A = \begin{pmatrix}
\frac{1}{4} & \frac{1}{4} & 0 & \frac{1}{4} \\
\frac{1}{4} & \frac{1}{4} & \frac{1}{2} & \frac{1}{4} \\
\frac{1}{4} & \frac{1}{4} & 0 & \frac{1}{4} \\
\frac{1}{4} & \frac{1}{4} & \frac{1}{2} & \frac{1}{4}
\end{pmatrix}.
$$

Then the probability distribution $\mathcal{P}_1$ of α_1 is given by

$$\mathcal{P}_1 = A \cdot \mathcal{P}_0 = \left(\tfrac{3}{16}, \tfrac{5}{16}, \tfrac{3}{16}, \tfrac{5}{16}\right). \tag{2.1}$$

Similarly, the probability distribution $\mathcal{P}_2$ of α_2 is given by

$$\mathcal{P}_2 = A \cdot \mathcal{P}_1 = \left(\tfrac{13}{64}, \tfrac{19}{64}, \tfrac{13}{64}, \tfrac{19}{64}\right). \tag{2.2}$$

In general, the probability distribution $\mathcal{P}_k$ of α_k is given by the formula $\mathcal{P}_k = A \cdot \mathcal{P}_{k-1}$.

Transient Analysis. We now show that $A^k \mathcal{P}_0$ gets exponentially close to the eigenvector $\pi = (\tfrac{1}{5}, \tfrac{3}{10}, \tfrac{1}{5}, \tfrac{3}{10})$ of A in a small number of steps independent of D and the value of α.

The eigenvalues of A are $\lambda_1 = 1$ and $\lambda_2 = -1/4$, with the other eigenvalues being zero. We diagonalize the matrix to obtain $\Lambda(1, -\tfrac{1}{4}, 0, 0) = P^{-1} A P$ where P as usual consists of eigenvectors of A.

Let the eigenvectors corresponding to λ_1 and λ_2 be π and π' respectively. The angle between these two eigenvectors is 78.69 degrees. Let q_1 and q_2 be the projections of $\mathcal{P}_0$ onto the one dimensional spaces spanned by π and π' respectively, and let $q' = \mathcal{P}_0 - q_1 - q_2$. Then $A^k \mathcal{P}_0 = A^k(q_1 + q_2 + q') = q_1 + \lambda_2^k q_2$, since $A_k q' = 0$. Since λ_2 is bounded away from 1, it follows that $\|A^k \mathcal{P}_0 - \pi\|$ drops exponentially fast in k. Thus our steady state eigenvector π will dominate the behavior of $A^k \mathcal{P}_0$ for even modest values of k.

Corollary 2.7 *On average, for random values of $\alpha \gg |x|$, the* NAF *representation of α has $2/3$ of its output digits equal to 0.*

Proof. By Theorem 2.6, out of every ten instances of α_k, we expect two to be 0 mod 4, three to be 1 mod 4, two to be 2 mod 4, and three to be 3 mod 4. Hence we produce on average two values of $g_D(\alpha_k)$ equal to $(0,0)$, three values of $g_D(\alpha_k)$ equal to $(0,1)$, two values of $g_D(\alpha_k)$ equal to 0, and three values of $g_D(\alpha_k)$ equal to $(0,x)$. Counting up the digits, we find that on average 12 out of the 18 output digits are equal to 0.

2.2   Generalizations

The techniques described above generalize readily to any digit set $D = \{0,1\} \cup X$ where X consists of $2^n - 1$ elements belonging to prescribed congruence classes mod 2^n. For example, using $n = 3$ we have been able to construct digit sets with proven 78% asymptotic sparsity (compared with 67% in Corollary 2.7 and 74% in Corollary 2.10). However, as a compromise between readability and generality, and also for space reasons, we limit our analysis here to the case of digit sets having five elements. We consider digit sets of the form $D = \{0, 1, x, y, z\}$ where $x \equiv 3 \pmod 8$, $y \equiv 5 \pmod 8$ and $z \equiv 7 \pmod 8$ are negative. The transition matrix in this case has the same largest and second largest eigenvalues as in the $\{0, 1, x\}$ case, with all other eigenvalues being 0. We emphasize that one has considerable freedom in the design of the transition matrix and that the choices given here merely represent a useful baseline.

Definition 2.8. *For a digit set D of the above form, let $f_D \colon \mathbb{N} \to \mathbb{N}$ and $g_D \colon \mathbb{N} \to D \cup (D \times D)$ be the functions defined by*

$$
f_D(n) = \begin{cases} n/8 & n \equiv 0 \pmod 8 \\ (n-1)/8 & n \equiv 1 \pmod 8 \\ n/2 & n \equiv 2 \pmod 8 \\ (n-x)/8 & n \equiv 3 \pmod 8 \\ n/4 & n \equiv 4 \pmod 8 \\ (n-y)/8 & n \equiv 5 \pmod 8 \\ (n-2x)/8 & n \equiv 6 \pmod 8 \\ (n-z)/8 & n \equiv 7 \pmod 8 \end{cases}, \qquad
g_D(n) = \begin{cases} (0,0,0) & n \equiv 0 \pmod 8 \\ (0,0,1) & n \equiv 1 \pmod 8 \\ 0 & n \equiv 2 \pmod 8 \\ (0,0,x) & n \equiv 3 \pmod 8 \\ (0,0) & n \equiv 4 \pmod 8 \\ (0,0,y) & n \equiv 5 \pmod 8 \\ (0,x,0) & n \equiv 6 \pmod 8 \\ (0,0,z) & n \equiv 7 \pmod 8 \end{cases}.
$$

Theorem 2.9 *Let $\alpha \in \mathbb{N}$ and $D = \{0,1,x,y,z\}$ as above. Any NAF representation $(b_\ell \cdots b_2 b_1 b_0)_2$ of $f_D(\alpha)$ yields a NAF representation $(b_\ell \cdots b_2 b_1 b_0 \mathbin{\|} g_D(\alpha))_2$ of α via concatenation. Furthermore, the the probability distribution of the congruence class of $f_D^k(\alpha) \bmod 8$, for random uniformly selected integers $\alpha \in [0, N]$, where $N \gg |\max(x,y,z)|$, converges to the vector $(\frac{1}{10}, \frac{7}{40}, \frac{1}{10}, \frac{1}{8}, \frac{1}{10}, \frac{7}{40}, \frac{1}{10}, \frac{1}{8})$ as $k \to \infty$, with error bounded in magnitude by an exponential in k.*

Proof. By hypothesis, the initial input α has probability distribution $\mathcal{P}_0 = (\frac{1}{8}, \frac{1}{8}, \frac{1}{8}, \frac{1}{8}, \frac{1}{8}, \frac{1}{8}, \frac{1}{8}, \frac{1}{8})$ over the congruence classes mod 8. The probability distribution $\mathcal{P}_1$ for $f_D(\alpha)$ is computed as follows:

- Assume that α is uniformly distributed in $[0, N]$.
- If $\alpha \equiv 0 \pmod 8$, then $\alpha_1 = \alpha/8$ is uniformly distributed mod 8.
- If $\alpha \equiv 1 \pmod 8$, then $\alpha_1 = \frac{\alpha-1}{8}$ is uniformly distributed mod 8.
- If $\alpha \equiv 2 \pmod 8$, then $\alpha_1 = \alpha/2$ is uniformly 1 or 5 mod 8.
- If $\alpha \equiv 3 \pmod 8$, then $\alpha_1 = \frac{\alpha-x}{8}$ is uniformly distributed mod 8.
- If $\alpha \equiv 4 \pmod 8$, then $\alpha_1 = \alpha/4$ is uniformly 1, 3, 5 or 7 mod 8.
- If $\alpha \equiv 5 \pmod 8$, then $\alpha_1 = \frac{\alpha-y}{8}$ is uniformly distributed mod 8.
- If $\alpha \equiv 6 \pmod 8$, then $\alpha_1 = \frac{\alpha-2x}{8}$ is uniformly distributed mod 8.
- If $\alpha \equiv 7 \pmod 8$, then $\alpha_1 = \frac{\alpha-z}{8}$ is uniformly distributed mod 8.

Denote by B the transition matrix

$$
B = \begin{pmatrix}
\frac{1}{8} & \frac{1}{8} & 0 & \frac{1}{8} & 0 & \frac{1}{8} & \frac{1}{8} & \frac{1}{8} \\
\frac{1}{8} & \frac{1}{8} & \frac{1}{2} & \frac{1}{8} & \frac{1}{4} & \frac{1}{8} & \frac{1}{8} & \frac{1}{8} \\
\frac{1}{8} & \frac{1}{8} & 0 & \frac{1}{8} & 0 & \frac{1}{8} & \frac{1}{8} & \frac{1}{8} \\
\frac{1}{8} & \frac{1}{8} & 0 & \frac{1}{8} & \frac{1}{4} & \frac{1}{8} & \frac{1}{8} & \frac{1}{8} \\
\frac{1}{8} & \frac{1}{8} & 0 & \frac{1}{8} & 0 & \frac{1}{8} & \frac{1}{8} & \frac{1}{8} \\
\frac{1}{8} & \frac{1}{8} & \frac{1}{2} & \frac{1}{8} & \frac{1}{4} & \frac{1}{8} & \frac{1}{8} & \frac{1}{8} \\
\frac{1}{8} & \frac{1}{8} & 0 & \frac{1}{8} & 0 & \frac{1}{8} & \frac{1}{8} & \frac{1}{8} \\
\frac{1}{8} & \frac{1}{8} & 0 & \frac{1}{8} & \frac{1}{4} & \frac{1}{8} & \frac{1}{8} & \frac{1}{8}
\end{pmatrix}.
$$

Then $P_1 = B \cdot P_0$, and as in the case of Theorem 2.6, the limit distribution of $\mathcal{P}_k = B^k P_0$ for $f_D^k(\alpha)$ converges exponentially rapidly to the eigenvector $\pi = (\frac{1}{10}, \frac{7}{40}, \frac{1}{10}, \frac{1}{8}, \frac{1}{10}, \frac{7}{40}, \frac{1}{10}, \frac{1}{8})$ of B.

Corollary 2.10 *On average, for random values of $\alpha \gg |x|$, the NAF representation of α has $20/27$ of its output digits equal to 0.*

Proof. Counting in the same manner as Corollary 2.7, we find that for every forty instances of α_k, on average 80 out of the 108 output digits are equal to 0.

$\alpha = 111101011000010010110001111001111000011000110001111101010101101000100\backslash$
$00010111101010101011101111000111111110101001011110010000010111001010001111\backslash$
$110101011101100001110000100011110001110101010110011$

$\alpha = z00000y0010010000z00x000x0000z00z00y00000z00y0000100z00000y00x0000x0\backslash$
$0001000100x000x0x000x0z00x00100y0000100y0000y000010000y00y00100x0000x00x\backslash$
$000000100000z00x00y00y00x0001000x0x000x00y0010010111$

for $X = \{0, 1, -709, -947, -913\}$

$\alpha = z00z000100x00000010010000010000y000x000000z0000y001001000x01000x0010\backslash$
$000000100y0000z000x0x00000x0001000y00y00100x00z000x0z00y0000x0000x000010\backslash$
$000x00x000x000000000z0010000100000y0010111010101101100111100000001$

for $X = \{0, 1, -152397797, -272310435, -132159113\}$

Fig. 1. Examples of randomized sparse representations of a fixed 192-bit integer α with respect to random digit sets $X = \{0, 1, x, y, z\}$. In each representation, the least significant digits are written on the left.

3   Empirical Results

We begin by describing the standard technique for implementing elliptic curve scalar multiplication using non-adjacent form representations. Let α be an integer having a NAF representation $\alpha = (a_k \cdots a_1 a_0)$ with respect to some digit set $D = \{0, 1, x, y, z, \ldots\}$. Compute the point xQ (and also yQ, zQ etc. if needed). The computation of xQ is very fast if $|x|$ is small, and even for large values of $|x|$ there are some protocols (such as ElGamal encryption) for which the point Q is fixed, in which case xQ may be precomputed and stored. One can then compute $\alpha Q = \sum 2^k(a_k Q)$ using the usual double and add formula except with xQ (resp. yQ, zQ) in place of Q whenever the representation of d contains an x (resp. y, z) term as opposed to a 1 term. The efficiency of this calculation depends in large part on the proportion of terms in the representation which are nonzero, since these are the terms that trigger addition operations in the standard double and add formula.

In order to make this algorithm practical for random digit sets, we need to allow the use of integer representations which lack the non-adjacency property, since not every integer has a NAF representation with respect to every digit set. Without this allowance, the algorithm $\mathcal{A}$ would enter into an infinite loop when presented with input values α that lack NAF representations. Our approach is to revert to standard binary representation whenever the algorithm $\mathcal{A}$ encounters an input value of size less than that of one of the digits in the digit set. In this case, the maximum possible length of the ensuing purely binary portion is $\ell :=$ $\log(\max\{|x|, |y|, |z|, \ldots\})$. Hence, for $\alpha \gg \max\{|x|, |y|, |z|, \ldots\}$, the statistical analysis of the previous section remains valid for the $1 - \frac{\ell}{\log|\alpha|}$ fraction of the digit string which comprises the vast majority of the representation of α.

Figure 1 contains examples of a 192-bit integer represented in random sparse format with respect to various digit sets $\{0, 1, x, y, z\}$. Figure 2 compares the measured performance of the randomized exponentiation algorithm versus signed binary exponentiation as well as standard double-and-add exponentiation. On

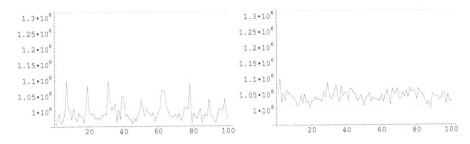

Fig. 2. Empirical timings for 192-bit EC exponentiation implemented on a 3.0 GHz Pentium 4 processor using the Microsoft bignum library. The vertical axis represents clock cycles and the horizontal axis depicts the results of 100 trials. In the left graph, each trial took place using a randomly selected digit set $\{0, 1, x, y, z\}$ with 10-bit values for x, y, z; the right graph uses 32-bit values. In each graph the two horizontal lines represent the cycle count for standard and signed binary representation, respectively.

average, the randomized algorithm outperforms signed binary multiplication for values of x, y, z as large as 10 bits, and remains competitive at 32-bit values. The timings do not include the cost of computing the individual multiples xQ, yQ, zQ, but in performance contexts this cost can be minimized by selecting small values for x, y, z. In the next section, however, we consider digit set randomization in the setting of side channel attacks, and in this setting we do need to use large values of x, y, z and account for the ensuing computational cost.

4   Digit Set Randomization as a Side Channel Attack Countermeasure

Side channel attacks [11] remain one of the most critical points of vulnerability for elliptic curve cryptosystem implementations as they exist today. These attacks make use of power consumption, cache hit rate, timing, or other differences between EC add and EC double operations to determine the binary representation of a scalar multiplier in an EC exponentiation operation [9]. While a number of protective countermeasures against side channel attacks have been proposed ([4,6,8,12,20,13,26,28]; see [3] for overview), many of the schemes have been broken [5,16,17,18,21,24,25,27] owing to their ad-hoc nature, and all of the existing proposals involve significant performance penalties.

We make a distinction between two classes of side channel attacks known as *simple* and *differential* attacks. In simple side channel attacks, a single execution instance is analyzed and the secret key is deduced using side channel information from that instance alone. In differential attacks, side channel information from multiple execution instances are compared and processed to deduce the secret key. For obvious reasons, it is generally considered more difficult to protect against differential attacks than against simple attacks. In this section we explain how randomized digit sets can be used to leverage simple side channel resilience into differential side channel resilience and present a formal analysis of security under a simplified attack model. Our goal here is not to provide a comprehensive proof of security, but rather just to suggest a new and promising type of approach which has never been considered before, and propose a preliminary naive security analysis as motivation.

A typical side channel attack operates by using side channel information to deduce the internal representation of a secret multiplier, for example by exploiting differences in power consumption between the main branches of a multiplication algorithm. In most cases, knowing the internal representation of an integer is enough to deduce the value of the multiplier. However, when randomized digit sets are used, a given internal (symbolic digit) representation can correspond to a multitude of different integer values, depending on which digit set is used. Hence, even if an attacker possesses full knowledge of the symbolic representation of an integer, we can still quantify to what extent does the value of the integer remain uncertain. Formally, we define the *open representation model* to denote the attack model in which the attacker possesses no information other than the symbolic digits corresponding to the secret multiplier α, and ask how many bits of information entropy remain in the value of α. In the next section we analyze this question and show that the number of bits is equal to the entropy of the digit set, assuming that this entropy is itself less than the entropy of α.

The computation of the individual multiples xQ, yQ, zQ in the RSF algorithm must be done in a side channel resistant manner in order to prevent the attacker from determining the values of x, y, z via side channel analysis. However, since x, y, z are randomly selected at runtime, the computation of xQ will only be performed once for any given value of x, and thus this computation only needs to resist simple side channel attacks.

Although some aspects of the open representation model lack realism—for example, a real attacker would likely know the value of αQ—we believe that the model is useful because it isolates the effects of side channel leakage in a well defined way. Our introduction of this attack model is novel since other side channel countermeasures do not make the crucial distinction between symbolic representations and integer values which is necessary in order for the model to be non-vacuous. Indeed, most side channel countermeasures in the literature rely on manipulating either the representation itself or the sequence of field operations used, and do not provide any security under the open representation model.

5   Entropy Bounds on Randomized Representations of Integers

In order to evaluate the security of digit set randomization in the open representation model, we now determine for a given digit string how many random digit sets D will produce a fixed number α under that digit string. In order to avoid the awkward issue of how to select random digit sets out of an infinite collection, we assume that the elements of D are bounded in absolute value by some fixed bound (such as 2^{32}) which is very small relative to α. We also assume for simplicity that the elements of D are all negative except for 0 and 1.

However, we emphasize that this analysis does not depend on the NAF property or indeed any other property of the digit string in question. Our analysis uses the Gaussian Heuristic [23] for lattices which states in any well behaved subset of $\mathbb{R}^n$ the number of lattice points inside is approximated by the ratio of the volume of the body to the lattice determinant.

5.1 One Random Term in D

The simplest case of randomized digit sets is sets of the form $D = \{0, 1\} \cup X$ where $X = \{x\}$, $x < 0$ is randomly selected (possibly under some mild constraints, such as $x \equiv 3 \pmod 4$, whose effect will be explained below). In this case, given a digit string $(a_k \cdots a_2 a_1 a_0)_2$, the corresponding value of α is

$$\alpha = \sum_{i=0}^{k} a_i 2^i = A_0 + A_1 x, \quad \text{where } A_0 = \sum_{a_i=1} 2^i, \quad A_1 = \sum_{a_i=x} 2^i.$$

For any given value of $\alpha \gg |x|$, there is only one value of x that will satisfy the equation $A_0 + A_1 x = \alpha$. Thus the information entropy of α is exactly equal to the entropy of X.

The condition $x \equiv 3 \pmod 4$ means that an attacker who obtains the complete representation of α can obtain the two least significant bits of α using the formula $\alpha = A_0 + A_1 x$. This phenomenon can also be seen in Figure 1 where the last two digits (or three digits, in the case of digit sets defined $\mod 8$) of the representation are independent of the digit set. However, this level of information leakage must be put into perspective: without digit set randomization, the entire integer α would already be known, as opposed to two or three bits.

5.2 Two Random Terms in D

If we consider digit sets $D = \{0, 1\} \cup X$ where $X = \{x, y\}$, then we have

$$\alpha = \sum_{i=0}^{k} a_i 2^i = A_0 + A_1 x + A_2 y, \quad \text{where } A_0 = \sum_{a_i=1} 2^i, A_1 = \sum_{a_i=x} 2^i, A_2 = \sum_{a_i=y} 2^i.$$

For fixed $\alpha, A_0, A_1, A_2 > 0$, the number of negative integer solutions (x, y) to $\alpha = A_0 + A_1 x + A_2 y$ (or, equivalently, the number of positive integer solutions (x, y) to $A_0 - \alpha = A_1 x + A_2 y$) is bounded above by

$$\frac{(A_0 - \alpha)}{A_1 A_2} \cdot \gcd(A_1, A_2).$$

This bound is obtained using standard linear Diophantine analysis. For convenience, we sketch the argument here. Let $Ax + By = C$ be a linear Diophantine equation in two variables, with $A, B, C > 0$. Divide out by $\gcd(A, B)$ to obtain a new equation $ax + by = c$ with $\gcd(a, b) = 1$. If (x_0, y_0) is one solution to the equation, then all solutions to the equation must have the form

$(x, y) = (x_0 + bt, y_0 - at)$, where t is an integer parameter. If we require x and y to be positive, then that imposes the bounds $0 < x < c/a$ on x, and the number of integers of the form $x = x_0 + bt$ that satisfy $0 < x < c/a$ is upper bounded by

$$\frac{c/a}{b} = \frac{c}{ab} = \frac{C}{AB} \cdot \gcd(A, B),$$

as desired. In particular, in expectation one would get $(A_0 - \alpha) = \Theta(\alpha) = \Theta(A_1) = \Theta(A_2)$ and $\gcd(A_1, A_2) = O(1)$, so $(A_0 - \alpha) \gcd(A_1, A_2) = \Theta(\alpha)$ is overwhelmingly likely to be less than $A_1 A_2 = \Theta(\alpha^2)$. Therefore, on average, we expect at most one negative integer solution (x, y) to the equation $\alpha = A_0 + A_1 x + A_2 y$, and thus the information entropy in computing α for the attacker is equal to the entropy in computing x and y.

5.3   General Case

In general, with $D = \{0, 1\} \cup X$, where $X = \{x_1, x_2, \ldots, x_c\}$, we find that the corresponding Diophantine equation $\alpha = A_0 + \sum_{i=1}^{c} A_i y_i$ has

$$\frac{1}{(c-1)!} \cdot \frac{(A_0 - \alpha)^{c-1}}{\prod_{i=1}^{c} A_i} \cdot \gcd(A_1, \ldots, A_c)$$

negative integer solutions. Here the numerator has approximate size $O(\alpha^{c-1})$ and the denominator $O(\alpha^c)$, so on average each α will have at most one negative integer solution.

6   Conclusions and Further Work

We present a method for using randomized digit sets in integer representations and give empirical results showing that elliptic curve point multiplication algorithms based on large randomized digit sets outperform both standard and signed binary representations. Our theoretical analysis of the sparsity of randomized digit set representations simplifies and generalizes the existing analyses available in the literature. We also propose digit set randomization as a side channel attack countermeasure, and provide a preliminary analysis of the security of random digit sets under a new attack model called the open representation model which is designed to isolate the impact of side channel information leakage. Our randomized algorithm is one of the only side channel countermeasures available that achieves even some level of security under this attack model.

In this paper we have not yet made any attempt to find parameters for digit set randomization which both simultaneously achieve good performance and good side channel resilience. In the future, we hope to perform empirical trials comparing the performance of random digit sets with various parameters against other existing side channel countermeasures; this task is greatly complicated by the large number and variety of side channel attack countermeasures which have been proposed. However, based on the fact that performance-oriented choices of digit set parameters lead to record or near record levels of performance, we are optimistic that digit set randomization provides a good foundation for future work towards high performing side channel attack resistant algorithms.

Acknowledgments. We are grateful to James Muir for his helpful and detailed suggestions to us during the preparation of this manuscript.

References

1. Avione, G., Monnerat, J., Peyrin, T.: Advances in Alternative Non-adjacent Form Representations. In: Canteaut, A., Viswanathan, K. (eds.) INDOCRYPT 2004. LNCS, vol. 3348, Springer, Heidelberg (2004)
2. Bosma, W.: Signed bits and fast exponentiation, 21st Journées Arithmétiques (English, with English and French summaries). J. Théor. Nombres Bordeaux 13(1), 27–41 (Rome, 2001)
3. Cohen, H., Frey, G.: Handbook of elliptic and hyperelliptic curve cryptography. In: Discrete Mathematics and its Applications(Boca Raton), Chapman & Hall/CRC, Boca Raton, FL (2006)
4. Coron, J.S.: Resistance against differential power analysis for elliptic curve. In: Koç, Ç.K., Paar, C. (eds.) CHES 1999. LNCS, vol. 1717, pp. 231–237. Springer, Heidelberg (1999)
5. Fouque, P.A., Muller, F., Poupard, G., Valette, F.: Defeating countermeasures based on randomized BSD representation. In: Joye, M., Quisquater, J.-J. (eds.) CHES 2004. LNCS, vol. 3156, pp. 312–327. Springer, Heidelberg (2004)
6. Ha, J.C., Moon, S.J.: Randomized Signed-Scalar Multiplication of ECC to Resist Power Attacks, In: Kaliski Jr., B.S., Koç, Ç.K., Paar, C. (eds.) CHES 2002. LNCS, vol. 2523, pp. 551–563. Springer, Heidelberg (2003)
7. Heuberger, C., Prodinger, H.: Analysis of alternative digit sets for nonadjacent representations. Monatsh. Math. 147(3), 219–248 (2006)
8. Itoh, K., Yajima, J., Takaneka, M., Torii, N.: DPA countermeasures by improving the window method. In: Kaliski Jr., B.S., Koç, Ç.K., Paar, C. (eds.) CHES 2002. LNCS, vol. 2523, pp. 303–317. Springer, Heidelberg (2003)
9. Jaffe, J., Jun, B., Kocher, P.: Differential Power Analysis. In: Wiener, M.J. (ed.) CRYPTO 1999. LNCS, vol. 1666, pp. 388–397. Springer, Heidelberg (1999)
10. Joye, M., Yen, S.-M.: Optimal left-to-right binary signed-digit recoding, IEEE Trans. On Computers 49(7), 740–748 (2000)
11. Kocher, P.: Timing attacks on implementations of Diffie-Hellman, RSA, DSS, and other systems. In: Koblitz, N. (ed.) CRYPTO 1996. LNCS, vol. 1109, Springer, Heidelberg (1996)
12. Liardet, P.-Y., Smart, N.P.: SPA/DPA in ECC systems using the Jacobi form. In: Koç, Ç.K., Naccache, D., Paar, C. (eds.) CHES 2001. LNCS, vol. 2162, Springer, Heidelberg (2001)
13. Möller, B.: Securing Elliptic Curve Point Multiplication against Side-Channel Attacks. In: Davida, G.I., Frankel, Y. (eds.) ISC 2001. LNCS, vol. 2200, Springer, Heidelberg (2001)
14. Morain, F., Olivos, J.: Speeding up the computations on an elliptic curve using addition-subtraction chains (English, with French summary). RAIRO Inform. Théor. Appl. 24(6), 531–543 (1990)
15. Muir, J.A., Stinson, D.R.: Alternative digit sets for nonadjacent representations. SIAM J. Discrete Math. 19(1), 165–191 (2005)

16. Okeya, K., Han, D.-G.: Side Channel Attack on Ha-Moon's Countermeasure of Randomized Signed Scalar Multiplication. In: Johansson, T., Maitra, S. (eds.) IN-DOCRYPT 2003. LNCS, vol. 2904, Springer, Heidelberg (2003)
17. Okeya, K., Sakurai, K.: A Second-Order DPA Attack Breaks a Window-method based Countermeasure against Side Channel Attacks. In: Chan, A.H., Gligor, V.D. (eds.) ISC 2002. LNCS, vol. 2433, Springer, Heidelberg (2002)
18. Okeya, K., Sakurai, K.: On Insecurity of the Side Channel Attack Countermeasure using Addition-Subtraction Chains under Distinguishability between Addition and doubling. In: Batten, L.M., Seberry, J. (eds.) ACISP 2002. LNCS, vol. 2384, Springer, Heidelberg (2002)
19. Osvik, D.A., Shamir, A., Tromer, E.: Cache Attacks and Countermeasures: The Case of AES (Extended Version). In: Park, C.-s., Chee, S. (eds.) ICISC 2004. LNCS, vol. 3506, Springer, Heidelberg (2005)
20. Oswald, E., Aigner, M.: Randomized addition-subtraction chains as a countermeasure against power attacks. In: Koç, Ç.K., Naccache, D., Paar, C. (eds.) CHES 2001. LNCS, vol. 2162, Springer, Heidelberg (2001)
21. Park, D.J., Lee, P.J.: A DPA Attack on the Improved Ha-Moon Algorithm. In: Song, J., Kwon, T., Yung, M. (eds.) WISA 2005. LNCS, vol. 3786, Springer, Heidelberg (2006)
22. Reitwiesner, G.W.: Binary arithmetic. Advances in computers 1, 231–308 (1960)
23. Schnorr, C.P., Hörner, H.H.: Attacking the Chor-Rivest cryptosystem by improved lattice reduction. In: Guillou, L.C., Quisquater, J.-J. (eds.) EUROCRYPT 1995. LNCS, vol. 921, Springer, Heidelberg (1995)
24. Sim, S.G., Park, D.J., Lee, P.J.: New power analyses on the Ha-Moon algorithm and the MIST algorithm. In: Lopez, J., Qing, S., Okamoto, E. (eds.) ICICS 2004. LNCS, vol. 3269, Springer, Heidelberg (2004)
25. Walter, C.D.: Breaking the Liardet-Smart randomized exponentiation algorithm, Smart Card Research and Advanced Applications—CARDIS, Usenix Association (2002)
26. Walter, C.D.: Some security aspects of the MIST randomzied exponentiation algorithm. In: Kaliski Jr., B.S., Koç, Ç.K., Paar, C. (eds.) CHES 2002. LNCS, vol. 2523, Springer, Heidelberg (2003)
27. Walter, C.D.: Issues of Security with the Oswald-Aigner Exponentiation Algorithm. In: Okamoto, T. (ed.) CT-RSA 2004. LNCS, vol. 2964, Springer, Heidelberg (2004)
28. Yen, S.M., Chen, C.N., Moon, S., Ha, J.: Improvement on Ha-Moon randomized exponentiation algorithm. In: Park, C.-s., Chee, S. (eds.) ICISC 2004. LNCS, vol. 3506, Springer, Heidelberg (2005)

Lower Bounds for Hit-and-Run Direct Search

Jens Jägersküpper*

Universität Dortmund, Informatik 2, 44221 Dortmund, Germany
JJ@Ls2.cs.uni-dortmund.de

Abstract. "Hit-and-run is fast and fun" to generate a random point in a high dimensional convex set K (Lovász/Vempala, MSR-TR-2003-05). More precisely, the hit-and-run random walk mixes fast independently of where it is started inside the convex set. To hit-and-run from a point $x \in \mathbb{R}^n$, a line L through x is randomly chosen (uniformly over all directions). Subsequently, the walk's next point is sampled from $L \cap K$ using a membership oracle which tells us whether a point is in K or not.

Here the focus is on black-box optimization, however, where the function $f \colon \mathbb{R}^n \to \mathbb{R}$ to be minimized is given as an oracle, namely a black box for f-evaluations. We obtain in an obvious way a direct-search method when we substitute the f-oracle for the K-membership oracle to do a line search over L, and, naturally, we are interested in how fast such a hit-and-run direct-search heuristic converges to the optimum point x^* in the search space $\mathbb{R}^n$.

We prove that, even under the assumption of perfect line search, the search converges (at best) linearly at an expected rate larger (i.e. worse) than $1 - 1/n$. This implies a lower bound of $0.5\,n$ on the expected number of line searches necessary to halve the approximation error. Moreover, we show that $0.4\,n$ line searches suffice to halve the approximation error only with an exponentially small probability of $\exp(-\Omega(n^{1/3}))$. Since each line search requires at least one query to the f-oracle, the lower bounds obtained hold also for the number of f-evaluations.

1 Introduction

Finding an optimum of a given function $f \colon S \to \mathbb{R}$ is one of the fundamental problems—in theory as well as in practice. The search space S can be discrete or continuous, like $\mathbb{N}$ or $\mathbb{R}$. If S has more than one dimension, it may also be a mixture. Here the optimization in "high-dimensional" Euclidean space is considered, i.e., the search space is $\mathbb{R}^n$. What "high-dimensional" means is usually anything but well defined. A particular 10-dimensional problem in practice may already be considered "high-dimensional" by the one who tries to solve it. Here the crucial aspect is how the optimization time scales with the dimensionality of the search space $\mathbb{R}^n$, i.e., we consider the optimization time as a function of n. In other words, here we are interested in what happens when the dimensionality

* Supported by the German Research Foundation (DFG) through the collaborative research center "Computational Intelligence" (SFB 531).

J. Hromkovič et al. (Eds.): SAGA 2007, LNCS 4665, pp. 118–129, 2007.
© Springer-Verlag Berlin Heidelberg 2007

of the search space gets higher and higher. This viewpoint is typical for analyses in computer science. In the domain of operations research and mathematical programming, however, focusing on how the optimization time scales with the search space's dimension seems not that common. Usually, the performance of an optimization method is described by means of convergence theory. As an example, let us take a closer look at "Q-linear convergence" (we drop the "Q" in the following): Let x^* denote the optimum search point of a unimodal function and $x^{[k]}$ the approximate solution after k optimization steps. Then we have

$$\frac{\text{dist}(x^*, x^{[k+1]})}{\text{dist}(x^*, x^{[k]})} \to r \in \mathbb{R}_{<1} \quad \text{as} \quad k \to \infty$$

where $\text{dist}(\cdot, \cdot)$ denotes some distance measure, most commonly the Euclidean distance between two points (when considering convergence towards x^* in the search space $\mathbb{R}^n$, as we do here), or the absolute difference in function value (when considering convergence towards the optimum function value in the objective space). Apparently, there seems to be no connection to n, the dimension of the search space. Yet only if r is an absolute constant, there is actual independence of n. In general, however, the *convergence rate* r depends on n. When we are interested in, say, the number of steps necessary to halve the approximation error (given by the distance from x^*), the order of this number with respect to n precisely depends on how r depends on n. For instance, if $r = 1 - 0.5/n$, we need $\Theta(n)$ steps; if $r = 1 - 0.5/n^2$, we need $\Theta(n^2)$ steps, and if $r = 1 - 2^{-n}$, we need $2^{\Theta(n)}$ steps. For any fixed dimension, however, in any of the three cases we actually have linear convergence. Thus, the order of convergence tells us something about the "speed" of the optimization, but in general nothing about the n-dependence of the number of steps necessary to ensure a certain approximation error (unless r is an absolute constant, i.e. independent of n). So, in case of linear convergence, we want to know how the convergence rate depends on the dimensionality of the search space.

Methods for solving optimization problems in continuous domains, essentially $S = \mathbb{R}^n$, are usually classified into first-order, second-order, and zeroth-order methods, depending on whether they utilize the gradient (first derivative) of the objective function, the gradient and the Hessian (second derivative), or neither of both. A zeroth-order method is also called *derivative-free* or *direct search*. Newton's method is a classical second-order method; first-order methods can be (sub)classified into Quasi-Newton, conjugate gradient, and steepest descent methods. Classical zeroth-order methods try to approximate the gradient and to then plug this estimate into a first-order method. Finally, amongst the modern zeroth-order methods, randomized direct-search heuristics like simulated annealing and evolutionary algorithms come into play, which are supposed general-purpose search heuristics.

When information about the gradient is not available, for instance if f relates to a property of some workpiece and is given by computer simulations or even by real-world experiments, then zeroth-order methods are the only option (unless simulations allow for algorithmic/automatic differentiation). As the approximation of the gradient usually involves at least n f-evaluations (forward finite differences; $2n$ for symmetric finite differences), a single optimization step of a classical zeroth order-method is computationally expensive, in particular if f is given implicitly by complex simulations. In practical optimization, especially in mechanical engineering, this is often the case, and particularly in this field randomized search heuristics (especially evolutionary algorithms) are becoming more and more popular. However, the enthusiasm in practical optimization heuristics has led to an unclear variety of very sophisticated and problem-specific algorithms. Unfortunately, from a theoretical point of view, the development of such algorithms is solely driven by practical success, whereas the aspect of a theoretical analysis is left aside.

In such situations f is given to the optimization algorithm as a black box for f-evaluations (zeroth-order oracle) and the cost of the optimization (the runtime) is defined as the number of queries to this oracle, and we are in the so-called *black-box optimization* scenario. Nemirovsky Yudin (1983, p. 333) state (w. r. t. optimization in continuous search spaces) in their book *Problem Complexity and Method Efficiency in Optimization:* "From a practical point of view this situation would seem to be more typical. At the same time it is objectively more complicated and it has been studied in a far less extent than the one [with first-order oracles/methods] considered earlier." After more than two decades there still seems to be some truth in their statement, though to a smaller extent. For discrete black-box optimization, a complexity theory has been successfully started, cf. Droste, Jansen, Wegener (2006). Lower bounds on the number of f-evaluations (the *black-box complexity*) are proved with respect to classes of functions when an arbitrary(!) optimization heuristic knows about the class $\mathcal{F}$ of functions to which f belongs, but nothing about f itself. The benefits of such results are obvious: They can prove that an allegedly poor performance of an apparently simple black-box algorithm on f is due to $\mathcal{F}$'s inherent black-box complexity rather than due to the algorithm's simpleness.

As already discussed above, the situation for heuristic optimization in continuous search spaces is different, especially with respect to randomized (direct) methods. The results to be presented here contribute to this emerging field of optimization theory.

2   The Framework for the Randomized Methods

As already noted above, classical zeroth-order methods (i. e. black-box optimizers) for continuous search spaces usually try to approximate the gradient of the function f to be minimized at the current search point x. Subsequently, a line search along gradient direction is performed to find the next search point, which replaces x. Usually, the line search aims at locating the best (with respect to the f-value) point on the line through x, and various strategies for how to do the line search exist (Armijo/Goldstein, Powell/Wolfe, etc.; cf. Nocedal Wright (2006, Ch. 3) for instance). As the approximation of the gradient usually involves at least n f-evaluations, and as the (approximate) gradient's direction may significantly differ from the direction pointing directly to the optimum x^*

anyway (cf. ill-conditioned quadratics), more and more direct-search heuristics have been proposed which abandon gradient approximation. Among the first and most prominent ones are the pattern search by Hooke Jeeves (1961) and the (downhill) simplex method by Nelder Mead (1965); cf. Kolda, Lewis, Torczon (2004) for a comprehensive review. Surprisingly, also already in the 1960s randomized direct-search methods were proposed, one is the so-called *evolution strategy* by Rechenberg (1965) and Schwefel (1965). For some obscure reason, however, there has been resentment against randomized algorithms in these early years. This started to change with the randomization of quicksort and randomized testing for primality. At the latest by the time when Dyer, Frieze, Kannan (1989) came up with a randomized approximation algorithm for the computation of the volume of a convex body in high dimensional space, the (potential) benefits of randomization have won recognition. Though the polynomial expected runtime of this algorithm was not very practical, it showed in principle the power of randomization since for any deterministic algorithm there is a convex set for which the relative approximation error is $n^{\Omega(n)}$ after any polynomial number of steps. At the core of this algorithm was a random walk on a (sufficiently fine) lattice. This algorithm was further improved, in particular by substituting the so-called *ball walk* for the original lattice walk. One step of this ball walk consists in uniformly choosing a point from the hyperball of radius δ around the current point. If this point lies in the convex set, then it becomes the next point of the walk. Obviously, one has to choose the parameter δ appropriately. Moreover, when the ball walk is started very close (w.r.t. δ) to the corner of a hypercube, just for instance, it may need an exponential number of steps to leave this corner, making a so-called warm start necessary (i.e. a preprocessing). As recently shown by Lovász Vempala (2006), using the *hit-and-run walk* instead of the ball walk avoids these two issues. Hit-and-run mixes fast even when started close to the boundary of the convex set, and moreover, no "step size" needs to be appropriately predefined. Also an optimization algorithm based on random walks in convex sets has been proposed (Bertsimas Vempala, 2004).

As already noted in the abstract, to hit-and-run from a point $x \in \mathbb{R}^n$ within a convex set $K \subset \mathbb{R}^n$, a line L through x is randomly chosen (uniformly over all directions). Subsequently, the walk's next point is sampled from $L \cap K$ (as uniformly as possible) using a membership oracle which tells us whether a sample from L lies in K or not. As also already noted in the abstract, we obtain in an obvious way a hit-and-run direct-search method for black-box optimization of $f: \mathbb{R}^n \to \mathbb{R}$ when we substitute the f-oracle for the K-membership oracle. Thus, the framework of the heuristics for black-box optimization we consider is as follows: For a given initialization of $x \in \mathbb{R}^n$ the following loop is performed:

1. Randomly choose a line L through x (uniformly over all directions).
2. By some kind of a line search (using the f-oracle), find a point $x' \in L$.
3. Set $x := x'$ and GOTO 1 (unless stopping is requested; then output x).

Naturally, we are interested in how fast such a heuristic converges to the optimum point $x^* \in \mathbb{R}^n$ (we assume that there is a unique global optimum), in particular:

How fast can it converge in principle? That is, we are interested in a general lower bound which is universal for the class of hit-and-run direct-search heuristics.

Note that there are *no* assumptions on how the line search is performed. In particular, for the line search in the ith iteration, the algorithm may use *all* the information gathered from *all* the samples drawn during the preceding $i-1$ line searches. Naturally, in each step the choice of how to do the line search may additionally depend on the actual direction of L. All in all, a large variety of adaptive strategies for black-box optimization is covered by our framework.

3   The Lower Bounds

Since any reasonable strategy for the line search implies at least one query to the f-oracle, in our scenario the number of f-evaluations is bounded below by the number of line searches. Thus, we focus on the number of line searches in the following and aim at a general lower bound. Therefore, we need an upper bound on the gain of a single line search. We consider the best case: When we want the heuristic to approach the unique optimum point x^* as fast as possible, we may optimistically assume that x' was chosen from the line L such that the distance between x' and x^* is minimum. Call this a *perfect line search*. The situation is depicted in the figure below.

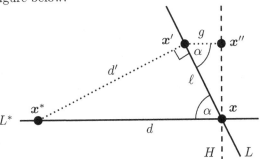

It is well known that the distance between x^* and x' is minimum when $x' \in L \supset \{x\}$ is such that the line passing through x' and x^* is perpendicular to the line L (given that $x^* \notin L$, which is the case with probability one, unlike already x coincides with the optimum point x^*, because L's direction is chosen uniformly over all directions).

Let $d := \mathrm{dist}(x, x^*)$ denote the current approximation error in the search space and let $d' := \mathrm{dist}(x', x^*)$. Furthermore, let L^* denote the line through x and x^*. Now consider the hyper-plane H which contains x and is perpendicular to L^*. Let $x'' := \arg\min_{y \in H} \mathrm{dist}(x', y)$ denote the unique point in H with smallest distance from x'. Then the angle α between L and L^* equals the angle between L and the line through x' and x'' (which is parallel to L^* since it is perpendicular to H just as L^*). Consequently, we have

$$d' = d \cdot \sin \alpha \qquad \text{and} \qquad \mathrm{dist}(x', H) = \mathrm{dist}(x', x) \cdot \cos \alpha.$$

Let $g := \text{dist}(\boldsymbol{x}', \boldsymbol{x}'')$ denote the distance of $\boldsymbol{x}'$ from H, and $\ell := \text{dist}(\boldsymbol{x}', \boldsymbol{x})$ so that we have $g/\ell = \cos\alpha$. Since $d'/d = \sin\alpha = \sqrt{1 - (\cos\alpha)^2}$, we obtain

$$\frac{d'}{d} = \sqrt{1 - (g/\ell)^2}, \tag{1}$$

which ranges in $[0, 1]$ since $g \in [0, \ell]$. Thus, instead of focusing on the distribution of $\sin\alpha$ when the line L is chosen uniformly over all directions, we can focus on the ratio g/ℓ and concentrate on the distribution of this *relative distance* of $\boldsymbol{x}'$ from the hyper-plane H (namely, relative to the distance of $\boldsymbol{x}'$ from $\boldsymbol{x}$). (It will shortly become clear why this makes sense.)

In two dimensions, like in the figure above, for any fixed $d' \in (0, d)$ there are exactly two (different) lines through $\boldsymbol{x}$ with distance d' from the optimum point $\boldsymbol{x}^*$. (Note that by fixing d' we also fixed ℓ and g.) In three or more dimensions, however, there is an infinite number of such lines. In three dimensions they form a double cone with its apex at $\boldsymbol{x}$, and all points of this cone with a distance of exactly d' from $\boldsymbol{x}^*$ (namely all $\boldsymbol{x}'$) form a circle. This circle lies in a plane which is parallel to H (a plane in three dimensions). In general, i.e. in $n \geq 3$ dimensions, the potential points $\boldsymbol{x}'$ form the set $S := \{\boldsymbol{x}' \in \mathbb{R}^n \mid \text{dist}(\boldsymbol{x}', \boldsymbol{x}^*) = d' \text{ and } \text{dist}(\boldsymbol{x}', \boldsymbol{x}) = \ell\}$, which is an $(n-1)$-sphere since S is the intersection of two hyper-spheres, namely of the hyper-sphere with radius d' centered at $\boldsymbol{x}^*$ and the hyper-sphere with radius ℓ centered at $\boldsymbol{x}$. Moreover, S lies in the hyper-plane H' which is parallel to H such that it has distance g from H and distance $d - g$ from $\boldsymbol{x}^*$. The situation is depicted below, where the left sphere consists of all points with distance d' from the optimum point $\boldsymbol{x}^*$, and the right sphere consists of all points with distance ℓ from our current approximate solution $\boldsymbol{x}$.

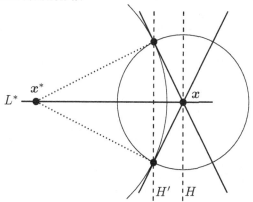

Recall that we fixed $d' \in (0, d)$ for the above discussion, and that this implies fixed values for ℓ and $g = \text{dist}(H', H)$. Now consider a randomly chosen line L through $\boldsymbol{x}$ (uniform over all directions). According to our construction, if L penetrates the $(n-1)$-sphere $S \subset H'$, then the perfect line search on L yields a point with a distance of exactly d' from $\boldsymbol{x}^*$. Now, if L lies inside the double cone, i.e., L penetrates the open $(n-1)$-ball whose missing boundary is S, then the perfect line search yields a point with a distance smaller than d' from $\boldsymbol{x}^*$. If L

lies outside the double cone (except for passing through the apex x, of course), then the perfect line search yields a point with a distance larger than d' from x^*. Thus, we are interested in the probability that L is chosen such that it lies inside the cone, namely the probability that the perfect line search yields a point with a distance of less than d' from x^*.

Now, how can we actually pick a line through x such that its direction is uniformly random? We pick uniformly at random a point y from/over the unit hyper-sphere centered at x and choose L as the line through y and x. From this point of view, the perfect line search yields a point with a distance of exactly d' from x^* if y's distance from H is exactly g/ℓ; a point with a distance smaller than d' from x^* if y's distance from H is larger than g/ℓ; and a point with a distance larger than d' from x^* if y's distance from H is smaller than g/ℓ.

In other words, we can consider the random variable $R := d'/d$ as a function of the random variable G defined as y's distance from the hyper-plane H, where the point y is chosen uniformly over the unit hyper-sphere centered at x. Namely, we have $R = \sqrt{1 - G^2}$, cf. Equation 1. (Note that the distribution of y over $\mathbb{R}^n$ is spherically symmetric; more precisely, invariant w. r. t. orthonormal transformations.) For $n \geq 4$ the density function of G's distribution over $[0, 1]$ is given by $(1 - x^2)^{(n-3)/2}/\Psi$ (Jägersküpper, 2003), where $\Psi = \int_0^1 (1 - x^2)^{(n-3)/2}\, dx$ (normalization) and the value of this integral is $\Psi = \sqrt{\pi/4} \cdot \Gamma(n/2 - 1/2)/\Gamma(n/2) = \sqrt{\pi/n}/2 + \Theta(n^{-3/2})$, where "$\Gamma$" denotes the well-known gamma function. Consequently, y's expected distance from H equals $\int_0^1 x \cdot (1 - x^2)^{(n-3)/2}\, dx/\Psi = (n - 1)^{-1}/\Psi$ which turns out to be about $0.8/\sqrt{n}$. This might appear bewildering (at first) since this implies that, as the search space's dimensionality increases, the expected distance from H tends to zero—although y's distance form x is fixed to one and H is hit with zero probability. However, noting that H is an affine subspace with dimension $n-1$ (i. e. codimension 1), it may become more plausible that getting far away from H becomes less and less probable as n increases. It might help even more to recall that an n-hypercube with unit diameter (longest diagonal) has edges of length $1/\sqrt{n}$.

Naturally, $\mathsf{E}[G]$ does not tell us much about $\mathsf{E}[R] = \mathsf{E}[\sqrt{1 - G^2}]$, the expectation in which we are actually interested. We can easily compute it, though:

$$\mathsf{E}[R] = \int_0^1 \sqrt{1 - x^2} \cdot (1 - x^2)^{(n-3)/2}/\Psi\, dx = \int_0^1 (1 - x^2)^{n/2-1}\, dx/\Psi.$$

Since $\int_0^1 (1 - x^2)^{n/2-1}\, dx = \sqrt{\pi/4} \cdot \Gamma(n/2)/\Gamma(n/2 + 1/2)$, we obtain

$$\mathsf{E}[R] = \frac{\sqrt{\pi/4} \cdot \Gamma(n/2)/\Gamma(n/2 + 1/2)}{\sqrt{\pi/4} \cdot \Gamma(n/2 - 1/2)/\Gamma(n/2)} = \frac{\Gamma(n/2) \cdot \Gamma(n/2)}{\Gamma(n/2 - 1/2) \cdot \Gamma(n/2 + 1/2)}. \tag{2}$$

Using $\Gamma(n/2 + 1/2) = \Gamma(n/2 - 1/2) \cdot (n/2 - 1/2)$, we have

$$\mathsf{E}[R] = \frac{n - 1}{2} \cdot \left(\frac{\Gamma(n/2)}{\Gamma(n/2 + 1/2)}\right)^2,$$

and since $\Gamma(n/2 + 1/2)/\Gamma(n/2) < \sqrt{n/2}$, we obtain the following lower bound on the expected factor by which the approximation error is reduced in each step:

$$\mathsf{E}[R] > \frac{n-1}{2} \cdot \frac{2}{n} = 1 - \frac{1}{n}.$$

This lower bound holds for perfect line search and, as a direct consequence, also for any other line-search strategy. With other words, this bound is universal for the class of hit-and-run direct-search methods.

To see how good this general lower bound on $\mathsf{E}[R]$ actually is, an upper bound on $\mathsf{E}[R]$ under the assumption of perfect line search would be nice. Using that $\Gamma(n) = (n-1)!$, $\Gamma(n/2) = (n-2)!! \cdot \sqrt{\pi}/2^{(n-1)/2}$, and $\Gamma(k+1/2) = (2k-1)!! \cdot \sqrt{\pi}/2^k$ (where $k!!$ is defined as $2 \cdot 4 \cdot 6 \cdots k$ for even k, and as $1 \cdot 3 \cdot 5 \cdots k$ for odd k), the right-hand side of Equation 2 can be estimated as follows:

$$\mathsf{E}[R] < \frac{2n-1}{2n} = 1 - \frac{1}{2n}.$$

In other words, for perfect line search, the expected factor by which the approximation error is reduced (in each step) is smaller (i. e. better) than $1 - 0.5/n$. This shows that our general lower bound of $1 - 1/n$ on $\mathsf{E}[R]$ is actually pretty tight. All in all, we have proved the following result:

Theorem 1. *Consider the optimization of a function $f \colon \mathbb{R}^n \to \mathbb{R}$ with a unique optimum point $\boldsymbol{x}^*$. Then we have for $n \geq 4$:*

The (hypothetical) hit-and-run direct-search method which performs a perfect line search in each step converges linearly towards $\boldsymbol{x}^*$ at an expected rate of $1 - \beta/n$, where $0.5 < \beta < 1$ (and β depends on n according to Equation 2).

Independently of how a hit-and-run direct-search method performs the line searches, the expected factor by which the approximation error (i. e. the distance from $\boldsymbol{x}^*$) is reduced is larger than $1 - 1/n$ in each step. That is, if (at all) a hit-and-run direct-search method converges towards $\boldsymbol{x}^*$, then at best linearly at an expected rate larger (i. e. worse) than $1 - 1/n$.

The result on the expected factor by which the approximation error is reduced directly implies a bound on the expected spatial gain towards the optimum point $\boldsymbol{x}^*$. Therefore, let $d^{[i]}$ denote the approximation error (i. e. the distance from $\boldsymbol{x}^*$) after the ith step, and let $d^{[0]}$ denote the initial approximation error. For a fixed $d^{[i-1]}$, let $\Delta^{[i]} := d^{[i]} - d^{[i-1]}$ be defined as the random variable corresponding to the spatial gain towards $\boldsymbol{x}^*$ in the ith step. Then the above theorem says that in general, i. e. for any hit-and-run direct-search method, $\mathsf{E}[\Delta^{[i]}] < d^{[i-i]}/n$ in each step i. Moreover, for perfect line search, in each step $\mathsf{E}[\Delta^{[i]}] = \beta(n) \cdot d^{[i-1]}/n$ for some function $\beta \colon \mathbb{N} \to (0.5, 1)$.

Let us stick with perfect line search in the following. Then the approximation error is non-increasing, i. e., $d^{[0]} \geq d^{[1]} \geq d^{[2]} \ldots$ (actually, $d^{[i+1]} < d^{[i]}$ with probability one, since the randomly chosen line lies in H with zero probability). Thus, in each step $\Delta^{[i]} < d^{[i-1]}/n \leq d^{[0]}/n$, and consequently, the number of steps necessary for an expected total gain of at least $d^{[0]}/2$ is larger than

$(d^{[0]}/2)/(d^{[0]}/n) = n/2$ (actually, $n \ln 2$). However, in general, maximizing the *expected total gain* of a fixed number of steps need not necessarily result in minimizing the *expected number of steps* to realize a specified gain. Nevertheless, $n/2$ will turn out to be a lower bound on the expected number of steps which are necessary to halve the approximation error. The proof is a straight-forward application of the following lemma, which is a modification of Wald's equation.

Lemma 2. *Let $X_1, X_2, \ldots$ denote random variables with bounded range and S the random variable defined by $S = \min\{ t \mid X_1 + \cdots + X_t \geq g\}$ for a given $g > 0$. Given that S is a stopping time (i.e., the event $\{S = t\}$ depends only on $X_1, \ldots, X_t$), if $\mathsf{E}[S] < \infty$ and $\mathsf{E}[X_i \mid S \geq i] \leq u \neq 0$ for all X_i, then $\mathsf{E}[S] \geq g/u$.*

(A proof can be found, e.g., in Jägersküpper, 2007.) We let X_i denote $\Delta^{[i]}$ and choose $g := d^{[0]}/2$. As we have just seen, $0 \leq \Delta^{[i]} \leq d^{[0]}$, and since in our scenario "$S \geq i$" merely means that the approximation error has not been halved in the first $i-1$ steps, actually $\mathsf{E}[\Delta^{[i]} \mid S \geq i] < d^{[0]}/n =: u$. Finally, we note that S is in fact a stopping time so that $g/u = n/2$ is indeed a lower bound on the expected number of steps to halve the approximation error (unless $\mathsf{E}[S]$ was infinite, in which case we would not need to prove a lower bound anyway). Due to the linearity of expectation, the expected number of steps to halve the approximation error $b \in \mathbb{N}$ times is lower bounded by $(n/2) + (b-1) \cdot (n/2 - 1)$, where the rightmost "$-1$" emerges because the last step within a halving-phase is also (and must be counted as) the first step of the following halving-phase. Thus, we have just proved the following result.

Theorem 3. *Let a hit-and-run direct-search method optimize a function in $\mathbb{R}^n$, $n \geq 4$, with a unique optimum. Let $b \colon \mathbb{N} \to \mathbb{N}$. For perfect line search, the expected number of steps until the approximation error in the search space is less than a $2^{-b(n)}$-fraction of the initial one is lower bounded by $b(n) \cdot n/2 - b(n) + 1$.*

Now that we know that at least $n/2$ steps are necessary *in expectation* to halve the approximation error, we would like to know whether there is a good chance of getting by with considerably fewer steps. In fact, we want to show that there is almost no chance of getting by with a little fewer steps. Actually, we are going to prove that $0.4\,n$ steps suffice to halve the approximation error only with an exponentially small probability. Therefore recall the following notions and notations, where X and Y denote random variables:

- X *stochastically dominates* Y, in short "$X \succ Y$," if (and only if) $\forall a \in \mathbb{R}$: $\mathsf{P}\{X \leq a\} \leq \mathsf{P}\{Y \leq a\}$. Obviously, "$\succ$" is a transitive relation.
- If $X \succ Y$ as well as $Y \succ X$, i.e., $\forall a \in \mathbb{R} \colon \mathsf{P}\{X \leq a\} = \mathsf{P}\{Y \leq a\}$, then X and Y are equidistributed and we write "$X \sim Y$."

Theorem 4. *Let a hit-and-run direct-search method optimize a function in $\mathbb{R}^n$ with a unique optimum. Let $b \colon \mathbb{N} \to \mathbb{N}$ such that $b(n) = \mathrm{poly}(n)$. For perfect line search, with a very high probability of $1 - \exp(-\Omega(n^{1/3}))$ more than $b(n) \cdot 0.4\,n$ steps are necessary until the approximation error is less than a $2^{-b(n)}$-fraction of the initial approximation error.*

Proof. Assume that $x^{[0]} \neq x^*$. Because in each step perfect line search is performed, $\Delta^{[i]}/d^{[i-1]} \sim \Delta^{[j]}/d^{[j-1]}$ for $i,j \in \mathbb{N}$ (scale invariance) . Since moreover $d^{[0]} \geq d^{[1]} \geq d^{[2]}\ldots$, we have $\Delta^{[1]} \succ \Delta^{[2]} \succ \ldots$ for the single-step gains. Let $X_1, X_2, X_3, \ldots$ denote independent instances of the random variable $\Delta^{[1]}$. Then $\forall i \in \mathbb{N}: X_i \succ \Delta^{[i]}$, and hence $\sum_{i=1}^{k} \Delta^{[i]} \prec S_k := \sum_{i=1}^{k} X_i$. In less formal words: Adding up k independent instances of the random variable which corresponds the spatial gain in the first step results in a random variable (namely S_k) which stochastically dominates the random variable given by the total gain of the first k steps. The advantage of considering S_k instead of the "true" total gain of these steps is the following: S_k is the sum of independent random variables so that we can apply Hoeffding's bound. Namely, Hoeffding (1963, Theorem 2) tells us:

Let $X_1, \ldots, X_k$ denote independent random variables with bounded ranges so that $a_i \leq X_i \leq b_i$ with $a_i < b_i$ for $i \in \{1, \ldots, k\}$. Let $S := X_1 + \cdots + X_k$. Then $\mathsf{P}\{S \geq \mathsf{E}[S] + x\} \leq \exp(-2x^2 / \sum_{i=1}^{k} (b_i - a_i)^2)$ for any $x > 0$.

If the support of each random variable X_i is contained in $[a, b] \subset \mathbb{R}$, Hoeffding's bound becomes $\exp(-2 \cdot (x/(b-a))^2/k)$. So, let $k := 0.4n$ and $S := S_k$. Then $\mathsf{E}[S] = 0.4n \cdot \mathsf{E}[\Delta^{[1]}] \leq 0.4d^{[0]}$, and for the application of Hoeffding's bound we choose $x := 0.1d^{[0]}$, which yields an upper bound of $\exp(-0.05(d^{[0]}/(b-a))^2/n)$ on the probability that the approximation error is halved in $0.4n$ steps. We can choose $a := 0$ so that we obtain $\mathsf{P}\{X_1 + \cdots + X_k \geq d^{[0]}/2 \mid X_1, \ldots, X_k \leq b\} \leq \exp(-0.05(d^{[0]}/b)^2/n)$, where b is an upper bound on the gain towards the optimum point x^* in a step. Unfortunately, when substituting the trivial upper bound of $d^{[0]}$ for b, the upper bound on the probability becomes $\exp(-0.05/n)$, which tends to one as n grows. For $b := d^{[0]}/n^{2/3}$, however, we obtain (recall that k was chosen as $0.4n$)

$$\mathsf{P}\left\{X_1 + \cdots + X_k \geq d^{[0]}/2 \mid X_1, \ldots, X_k \leq d^{[0]}/n^{2/3}\right\} \leq e^{-0.05\,n^{1/3}}.$$

Thus, if we can show that $\mathsf{P}\{X_i > d^{[0]}/n^{2/3}\} = e^{-\Omega(n^{1/3})}$ in each of the $0.4n$ steps, we obtain (by an application of the union bound)

$$\mathsf{P}\left\{X_1 + \cdots + X_k \geq d^{[0]}/2\right\} \leq e^{-0.05\,n^{1/3}} + 0.4n \cdot e^{-\Omega(n^{1/3})} = e^{-\Omega(n^{1/3})}.$$

Finally, by another application of the union bound, we obtain the theorem because $b(n) = \text{poly}(n)$ implies $b(n) \cdot e^{-\Omega(n^{1/3})} = e^{-\Omega(n^{1/3})}$.

In other words, it remains to be shown that $\mathsf{P}\{\Delta^{[0]} > d^{[0]}/n^{2/3}\}$ is actually bounded above by $e^{-\Omega(n^{1/3})}$. Therefore, recall Equation 1. It tells us that $d - d' = d \cdot (1 - \sqrt{1 - (g/\ell)^2})$. As a consequence, $\mathsf{P}\{\Delta > d/n^{2/3}\}$ is equal to $\mathsf{P}\{1 - \sqrt{1 - G^2} > 1/n^{2/3}\}$. Solving the inequality $1 - \sqrt{1 - G^2} > 1/n^{2/3}$ for G yields $G > \sqrt{2/n^{2/3} + 1/n^{4/3}}$ so that that $\Delta > d/n^{2/3}$ actually implies $G > \sqrt{2}/n^{1/3}$. Since G's density is a non-increasing function in $[0, 1]$,

$$\mathsf{P}\left\{G > \sqrt{2}/n^{1/3}\right\} = \int_{\sqrt{2}/n^{1/3}}^{1} (1 - x^2)^{(n-3)/2}\,\mathrm{d}x < \left(1 - \frac{2}{n^{2/3}}\right)^{(n-3)/2}.$$

Since $(1-t/k)^k < e^{-t}$ for $0 < t < k > 1$, we have $(1 - 2/n^{2/3})^{n^{2/3}} < e^{-2}$, so that $\mathsf{P}\{\Delta > d/n^{2/3}\} < \mathsf{P}\{G > \sqrt{2}/n^{1/3}\} < e^{-2\cdot((n-3)/2)/n^{2/3}} = e^{-n^{1/3}+3/n^{2/3}}$.   □

4 Discussion and Conclusion

Even though it is clear from intuition that the lower bounds presented in the two preceding theorems do not only hold for perfect line search but for any line-search strategy, they are formally proved only for perfect line search. Interestingly, we can easily show that our theorems do hold independently of how the line searching is actually done: By induction over the number of steps i we show that the random variable which corresponds to the approximation error after i steps for a given line-search strategy stochastically dominates the random variable $d^{[i]}$ for perfect line search, which we considered in the proofs.

So, hit-and-run direct-search methods converge (at best and if at all) linearly with an expected rate larger/worse than $1 - 1/n$. In simple words, the reason for this is that in high dimensions the randomly chosen direction is with a high probability "almost perpendicular" to the direction pointing directly towards the optimum point $\boldsymbol{x}^*$. For the further discussion, consider the simple toy problem of minimizing a quadratic form $\boldsymbol{x} \mapsto \boldsymbol{x}^\top \boldsymbol{Q}\boldsymbol{x}$, where the matrix $\boldsymbol{Q} \in \mathbb{R}^{n\times n}$ is positive definite. For this simple scenario, steepest descent converges at least linearly at a rate which is independent of the dimension n but which gets worse when the condition number of $\boldsymbol{Q}$ increases—when assuming a worst-case starting point (cf. Nocedal Wright (2006, Sec. 3.3) for instance). In the best case, however, steepest descent needs a single (perfect) line search to determine the optimum. Thus, for ill-conditioned quadratics, the performance of steepest decent heavily depends on the starting point. This is one reason why usually preconditioning is applied. Hypothetically assume for a moment the extreme of perfect preconditioning, so that $\boldsymbol{x}^\top \boldsymbol{I}\boldsymbol{x} = |\boldsymbol{x}|^2$ is to be minimized. Interestingly, the original evolution strategy from 1965 by Rechenberg/Schwefel mentioned in the introduction, a very simple randomized method which belongs to the class of hit-and-run direct-search methods (a line search consists in sampling a single point), actually gets by with $O(n)$ f-evaluations with very high probability to halve the approximation error in this scenario (Jägersküpper, 2003). This shows that the very general lower bound obtained here can be met at least up to a constant factor. However, in this ideal scenario steepest descent needs a single (perfect) line search to find the optimum independently of the starting point. Now, as we consider black-box optimization, steepest descent must approximate the gradient. Even though the approximation of the gradient may cost $2n$ f-evaluations, a single line search in this approximate direction may yield a significantly larger gain towards the optimum—whereas a hit-and-run method needs at least $0.5n$ f-evaluations to halve the approximation error *in any case* (in expectation; $0.4n$ with very high probability). Thus, with a passable preconditioning, the approximation of the gradient should pay off—even though it costs a linear (in n) number of f-evaluations per step—so that it will likely be superior to hit-and-run into a random direction.

As just discussed, for smooth functions we cannot expect hit-and-run direct search to compete with methods which learn (and then utilize) second-order information like the well-known BFGS method or (nonlinear) conjugate gradient methods. Clearly, hit-and-run *can* make sense in real-world optimization—when classical/established (nonlinear) methods have turned out to fail. For instance, when the function to be optimized is non-smooth, or disturbed by noise, or highly multimodal such that gradient approximation is deceptive. Then, however, as we have proved here, we should not expect such hit-and-run direct search to be fast.

References

Bertsimas, D., Vempala, S.: Solving convex programs by random walks. Journal of the ACM 51(4), 540–556 (2004)

Droste, S., Jansen, T., Wegener, I.: Upper and lower bounds for randomized search heuristics in black-box optimization. Theory of Computing Systems 39(4), 525–544 (2006)

Dyer, M.E., Frieze, A.M., Kannan, R.: A random polynomial time algorithm for approximating the volume of convex bodies. In: Proceedings of the 21st Annual ACM Symposium on Theory of Computing (STOC), pp. 375–381. ACM Press, New York (1989)

Fogel, D.B. (ed.): Evolutionary Computation: The Fossil Record. Wiley-IEEE Press, Chichester (1998)

Hoeffding, W.: Probability inequalities for sums of bounded random variables. American Statistical Association Journal 58(301), 13–30 (1963)

Hooke, R., Jeeves, T.A.: "Direct search" solution of numerical and statistical problems. Journal of the ACM 8(2), 212–229 (1961)

Jägersküpper, J.: Analysis of a simple evolutionary algorithm for minimization in Euclidean spaces. In: Baeten, J.C.M., Lenstra, J.K., Parrow, J., Woeginger, G.J. (eds.) ICALP 2003. LNCS, vol. 2719, Springer, Heidelberg (2003)

Jägersküpper, J.: Algorithmic analysis of a basic evolutionary algorithm for continuous optimization. Theoretical Computer Science 379(3), 329–347 (2007)

Kolda, T.G., Lewis, R.M., Torczon, V.: Optimization by direct search: New perspectives on some classical and modern methods. SIAM Review 45(3), 385–482 (2004)

Lovász, L., Vempala, S.: Hit-and-run from a corner. SIAM Journal on Computing 35(4), 985–1005 (2006)

Nelder, J.A., Mead, R.: A simplex method for function minimization. The Computer Journal 7, 308–313 (1965)

Nemirovsky, A.S., Yudin, D.B.: Problem Complexity and Method Efficiency in Optimization. Wiley, Chichester (1983)

Nocedal, J., Wright, S.: Numerical Optimization, 2nd edn. Springer, Heidelberg (2006)

Rechenberg, I.: Cybernetic solution path of an experimental problem. Royal Aircraft Establishment. In: Fogel (ed.) (1965)

Schwefel, H.-P: Kybernetische Evolution als Strategie der experimentellen Forschung in der Strömungstechnik. Diploma thesis, Technische Universität Berlin (1965)

An Exponential Gap Between LasVegas and Deterministic Sweeping Finite Automata*

Christos Kapoutsis, Richard Královič, and Tobias Mömke

Department of Computer Science, ETH Zürich

Abstract. A two-way finite automaton is *sweeping* if its input head can change direction only on the end-markers. For each $n \geq 2$, we exhibit a problem that can be solved by a $O(n^2)$-state sweeping *LasVegas* automaton, but needs $2^{\Omega(n)}$ states on every sweeping *deterministic* automaton.

1 Introduction

One of the major goals of the theory of computation is the comparative study of probabilistic computations, on one hand, and deterministic and nondeterministic computations, on the other. An important special case of this comparison concerns probabilistic computations of zero error (also known as "LasVegas computations"): how does ZPP compare with P and NP? Or, in informal terms: *Can every fast LasVegas algorithm be simulated by a fast deterministic one? Can every fast nondeterministic algorithm be simulated by a fast LasVegas one?*

Naturally, the computational model and resource for which we pose these questions are the Turing machine and time, respectively, as these give rise to the best available theoretical model for the practical problems that we care about. However, the questions have also been asked for other computational models and resources. Of particular interest to us is the case of restricted models, where the questions appear to be much more tractable. Conceivably, answering them there might also improve our understanding of the harder, more general settings.

In this direction, Hromkovič and Schnitger [1] studied the case of one-way finite automata, where efficiency is measured by size (number of states). They showed that, in this context, LasVegas computations are not more powerful than deterministic ones—intuitively, *every small one-way LasVegas finite automaton* (1P₀FA) *can be simulated by a small deterministic one* (1DFA). This immediately implied that, in contrast, nondeterministic computations are more powerful than LasVegas ones: *there exist small one-way nondeterministic finite automata* (1NFAs) *that cannot be simulated by any small* 1P₀FA.

For the case of two-way finite automata (2DFAs, 2P₀FAs, and 2NFAs), though, the analogous questions remain open [2]: *Can every small* 2P₀FA *be simulated by a small* 2DFA*? Can every small* 2NFA *be simulated by a small* 2P₀FA*?* Note that a negative answer to either question would confirm the long-standing conjecture that 2NFAs can be exponentially more succinct than 2DFAs [5].

---

* Work supported by the Swiss National Science Foundation grant 200021-107327/1.

J. Hromkovič et al. (Eds.): SAGA 2007, LNCS 4665, pp. 130–141, 2007.
© Springer-Verlag Berlin Heidelberg 2007

In this article we provide such negative answers for the special case where the two-way automata involved are *sweeping* (SDFAs, SP₀FAs, SNFAs), in the sense that their input head can change direction only on the end-markers. Both answers use the crucial fact (adapted from [2,4]) that a problem can be solved by small SP₀FAs iff small SNFAs can solve both that problem and its complement. Based on that, the answer to the second question is an immediate corollary of the recent result of [3]. The first question is answered by exhibiting a specific problem (inspired by *liveness* [5]) that cannot be solved by small SDFAs but is such that small SNFAs can solve both it and its complement. Our contribution is this latter theorem.

We stress that the expected running time of all probabilistic automata in this article is (required to be finite, but) allowed to be exponential in the length of the input, as our focus is on size complexity only. Our theorem should be interpreted as a first step towards the more natural (and more faithful to the analogy with ZPP, P, and NP) case where size and time must be held small simultaneously.

The next section defines the basics and presents the problem witnessing the separation. Section 3 describes a SP₀FA that solves this problem with $O(n^2)$ states. Section 4 proves that every SDFA solving the same problem needs at least $2^{\Omega(n)}$ states. Finally, Section 5 sketches a bigger picture that our theorem fits in.

2 Preliminaries

By $[n]$ we denote $\{1, 2, \ldots, n\}$. If Σ is an alphabet, then Σ^* is the set of all finite strings over Σ. If $z \in \Sigma^*$, then $|z|$, z_t, z^t, and z^R are its length, t-th symbol (if $1 \le t \le |z|$), t-fold concatenation with itself (if $t \ge 0$), and reverse. A *problem* (or *language*) over Σ is any $L \subseteq \Sigma^*$; then $\overline{L}$ is its complement. If $\# \notin \Sigma$, then $L^\#$ is the problem $\#(L\#)^*$ of all #-delimited finite concatenations of strings of L.

An automaton *solves* (or *recognizes*) a problem iff it accepts exactly the strings of that problem. A *family of automata* $M = (M_n)_{n \ge 0}$ solves a *family of problems* $\Pi = (\Pi_n)_{n \ge 0}$ iff, for all n, M_n solves Π_n. The automata of M are 'small' iff, for some polynomial p and all n, M_n has at most $p(n)$ states. Often, the generic member of a family informally denotes the family itself: e.g., "Π_n can be solved by a small 1DFA" means that some family of small 1DFAs solves Π.

If f is a function and $t \ge 1$, then f^t is the t-fold composition of f with itself.

Sweeping automata. A *sweeping deterministic finite automaton* (SDFA) [6] over an alphabet Σ and a set of states Q is any triple $M = (q_s, \delta, q_a)$ of a *start* state $q_s \in Q$, an *accept* state $q_a \in Q$, and a *transition function* δ which partially maps $Q \times (\Sigma \cup \{\vdash, \dashv\})$ to Q, for some *end-markers* $\vdash, \dashv \notin \Sigma$. An input $z \in \Sigma^*$ is presented to M surrounded by the end-markers, as $\vdash z \dashv$. The computation starts at q_s and on $\vdash$. The next state is always derived from δ and the current state and symbol. The next position is always the adjacent one in the direction of motion; except when the current symbol is $\vdash$ or when the current symbol is $\dashv$ and the next state is not q_a, in which cases the next position is the adjacent one towards the other end-marker. Note that the computation can either loop, or hang, or fall off $\dashv$ into q_a. In this last case we call it *accepting* and say that M *accepts* z.

More generally, for any input string $z \in \Sigma^*$ and state p, the *left computation of M from p on z* is the unique sequence $\text{LCOMP}_{M,p}(z) := (q_t)_{1 \leq t \leq m}$, where $q_1 := p$; every next state is $q_{t+1} := \delta(q_t, z_t)$, provided that $t \leq |z|$ and the value of δ is defined; and m is the first t for which this provision fails. If $m = |z| + 1$, we say that the computation *exits z into q_m*; otherwise, $1 \leq m \leq |z|$ and the computation *hangs at q_m*. The *right computation of M from p on z* is denoted by $\text{RCOMP}_{M,p}(z)$ and defined symmetrically, with $q_{t+1} := \delta(q_t, z_{|z|+1-t})$.

The *traversals of M on z* are the members of the unique sequence $(c_t)_{1 \leq t < m}$ where $c_1 := \text{LCOMP}_{M,p_1}(z)$ for $p_1 := \delta(q_s, \vdash)$; every next traversal c_{t+1} is either $\text{RCOMP}_{M,p_{t+1}}(z)$, if t is odd and c_t exits into a state q_t such that $\delta(q_t, \dashv) = p_{t+1} \neq q_a$, or $\text{LCOMP}_{M,p_{t+1}}(z)$, if t is even and c_t exits into a state q_t such that $\delta(q_t, \vdash) = p_{t+1}$; and m is either the first t for which c_t cannot be defined or ∞, if c_t exists for all t. Then, the *computation of M on z*, denoted by $\text{COMP}_M(z)$, is the concatenation of $(q_s), c_1, c_2, \ldots$ and possibly also (q_a), if m is finite and even and c_{m-1} exits into a state q_{m-1} such that $\delta(q_{m-1}, \dashv) = q_a$.

If M is allowed more than one next move at each step, we say it is *nondeterministic* (a SNFA). Formally, this means that δ partially maps $Q \times (\Sigma \cup \{\vdash, \dashv\})$ to the set of *all non-empty subsets* of Q. Hence, on any $z \in \Sigma^*$, $\text{COMP}_M(z)$ is a *set* of computations. If at least one of them is accepting, we say that M *accepts z*.

If M follows exactly one of its nondeterministic choices at each step according to some rational distribution, we say it is *probabilistic* (a SPFA). Formally, this means that δ partially maps $Q \times (\Sigma \cup \{\vdash, \dashv\})$ to the set of *all rational distributions* over Q—i.e., all total functions from Q to the rational numbers that obey the axioms of probability. Hence, on any $z \in \Sigma^*$, $\text{COMP}_M(z)$ is a *rational distribution* of computations. The expected length of a computation drawn from this distribution is called the *expected running time* of M on z.

For M to be a *Las Vegas* SPFA (a SP$_0$FA), a few extra conditions should hold. First, a special *reject* state $q_r \in Q$ must be specified—so that $M = (q_s, \delta, q_a, q_r)$. Second, whenever the current symbol is $\dashv$ and the next state is q_r, the next position is the adjacent one in the direction of motion—so that a computation may also fall off $\dashv$ into q_r, in which case we call it *rejecting*. Last, on any $z \in \Sigma^*$, a computation drawn from $\text{COMP}_M(z)$ must be either accepting with probability 1 or rejecting with probability 1. In the former case, we say that M *accepts z*. The concept of Las Vegas randomness is closely related to the self-verifying nondeterminism (see [2]).

Finally, a sweeping automaton is called *one-way* (1DFA, 1NFA, 1PFA, 1P$_0$FA) if it halts immediately after reading the right end-marker. Formally, this means that the value of the transition function on any state and on $\dashv$ is always either undefined or q_a (for 1DFAs); or $\{q_a\}$ (for 1NFAs); or the unique distribution over $\{q_a\}$ (for 1PFAs); or some distribution over $\{q_a, q_r\}$ (for 1P$_0$FAs).

The witness. In this section we define the family of problems Π that witnesses the separation between small SP$_0$FAs and small SDFAs. Let $n \geq 2$ be arbitrary.

Problem Π_n consists of all #-delimited concatenations of the strings of another problem, Π'_n. That is, $\Pi_n := (\Pi'_n)^\# = \#(\Pi'_n\#)^*$. So, we need to present Π'_n.

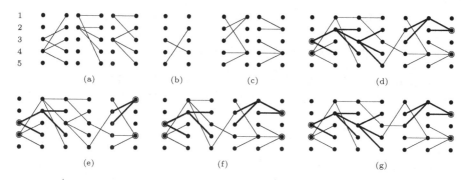

Fig. 1. (a) Three symbols of Γ_5; e.g., the leftmost one is $(3, 4, \{2, 4\})$. (b) The symbol $\{(3, 4), (5, 2)\}$ of X_5. (c) Two symbols of Δ_5. (d) The string defined by the six symbols of (a)-(c); in circles: the roots of the four trees; in bold: the two upper trees; the string is in Π'_5. (e) The upper left tree vanishes. (f) No tree vanishes, but the middle edges miss the upper left tree. (g) A well-formed string that does not respect the tree order.

Problem Π'_n is defined over the alphabet $\Sigma'_n := \Gamma_n \cup X_n \cup \Delta_n$, where:

$$\Gamma_n := \{ \, (i, j, \alpha) \mid i, j \in [n] \text{ and } i < j \text{ and } \emptyset \neq \alpha \subsetneq [n] \, \},$$
$$X_n := \{ \, \{(i, r), (j, s)\} \mid i, j, r, s \in [n] \text{ and } i \neq j \text{ and } r \neq s \, \},$$
$$\Delta_n := \{ \, (\alpha, j, i) \mid i, j \in [n] \text{ and } i < j \text{ and } \emptyset \neq \alpha \subsetneq [n] \, \}.$$

Intuitively, each $(i, j, \alpha) \in \Gamma_n$ represents a two-column graph (Fig. 1a) that has n nodes per column and contains exactly the edges that connect the ith left node to all right nodes inside α and the jth left node to all right nodes outside α. Symmetrically, each $(\alpha, j, i) \in \Delta_n$ represents a similar graph (Fig. 1c) containing exactly the edges that connect the ith and jth right nodes to the left nodes inside α and outside α, respectively. Finally, each $\{(i, r), (j, s)\} \in X_n$ represents a graph (Fig. 1b) containing only the edges connecting the ith and jth left nodes to the rth and sth right nodes, respectively. In all cases, we say that i and j (and r and s, in the last case) are the *roots* of the given symbol.

Of all strings over Σ'_n, consider those following the pattern $\Gamma_n^* X_n \Delta_n^*$. Each of them represents the multi-column graph (Fig. 1d) that we get from the corresponding sequence of two-column graphs when we identify adjacent columns. The symbol of X_n is called 'the middle symbol'—although it may very well not be in the middle position. If we momentarily hide the edges of that symbol, we easily see that the graph consists of exactly four disjoint trees, stemming out of the roots of the leftmost and rightmost columns. The tree out of the upper root of the leftmost column is naturally referred to as "the upper left tree". Similarly, the other trees are called "lower left", "upper right", and "lower right". Notice that, starting from the leftmost column, the two left trees may or may not both reach the left column of the middle symbol, as one of them may at some point 'cover all nodes' (Fig. 1e). Similarly, at least one of the two right trees reaches the right column of the middle symbol, but not necessarily both. Also observe that, in the case where all four trees make it to the middle symbol, the two edges of that symbol may or may not collectively 'touch' all trees (Fig. 1f). A string over Σ'_n is called *well-formed* if it belongs to $\Gamma^*_n X_n \Delta^*_n$ and is such that each of the four trees contains exactly one of the roots of the middle symbol (Fig. 1dg).

Of all well-formed strings over Σ'_n, problem Π'_n consists of those that 'respect the tree order', in the sense that the two edges of the middle symbol do not connect an upper tree to a lower one (Fig. 1d). In other words, this is the set

$$\Pi'_n := \{z \in (\Sigma'_n)^* \mid z \text{ is well-formed and respects the tree order}\}.$$

Hence, to solve $\Pi_n = \#(\Pi'_n\#)^*$ means to check that the input string (over $\Sigma_n := \Sigma'_n \cup \{\#\}$) starts and ends with # and is such that *every* infix between two successive copies of # is well-formed and respects the tree order.

3   The Upper Bound

In this section we prove that Π_n can be solved by a SP₀FA with $O(n^2)$ states.

One-way nondeterministic finite automata. The next two simple lemmata reduce solving Π_n with a small SP₀FA to solving Π'_n and $\overline{\Pi'_n}$ with small 1NFAs.

Lemma 1 (adapted from [2,4]). *If each of L and $\overline{L}$ can be solved by a 1NFA with m states, then L can be solved by a SP₀FA with $1 + 2m$ states.*

Proof. Suppose M and $\overline{M}$ are two m-state 1NFAs solving L and $\overline{L}$, respectively. Then, on any input z, exactly one of the computation trees of M and $\overline{M}$ on z contains accepting computations. We construct a SP₀FA M' for L that navigates probabilistically through these trees, trying to discover such a computation. If it succeeds, then it accepts or rejects, depending on which tree the computation was found in. If it fails, it sweeps back to the left end-marker and tries again.

More specifically, on input z, M' performs a series of sweeps. Each left-to-right sweep is an attempt to find an accepting computation of either M or $\overline{M}$ on z, while right-to-left sweeps are just rewinds. A left-to-right sweep starts with M' selecting one of M and $\overline{M}$ uniformly at random. Then, the selected 1NFA is simulated on z: at each step, M' either follows one of the possible next states uniformly at random or—if there are no such states (i.e., the 1NFA would hang at that point)—simply stops the simulation and sweeps blindly to $\dashv$. If the simulation ever reaches a situation where the 1NFA would be about to fall off $\dashv$ into its accepting state, then M' has discovered the desired accepting computation and therefore falls off $\dashv$, too, into its own accepting or rejecting state (depending on whether it had been simulating M or $\overline{M}$, respectively). Otherwise, the simulation stops somewhere before or at $\dashv$, in which case M' finishes the left-to-right sweep, sweeps back to $\vdash$, and starts a new attempt.

It is not hard to see that M' can be constructed out of a copy of M, a copy of $\overline{M}$, and 1 extra state. Also, M' halts only after finding an accepting computation, which happens with probability 1, and then decides correctly. Finally, since each attempt uses at most $2|z| + 2$ steps and succeeds with probability at least $\frac{1}{2}(\frac{1}{m})^{|z|+1}$, the average running time is at most $(2|z| + 2) \cdot 2m^{|z|+1} = 2^{O(|z|)}$.   □

Lemma 2. *If L can be solved by a 1NFA with m states, then $L^{\#}$ can be solved by an 1NFA with $2 + m$ states. Similarly, if $\overline{L}$ can be solved by a 1NFA with m states, then $\overline{L^{\#}}$ can be solved by an 1NFA with $4 + m$ states.*

Proof. Suppose M is an m-state 1NFA solving L. A 1NFA M' for $L^{\#}$ can simply simulate M successively on every #-delimited infix of its input, until the input is exhausted or one of these simulations produces no accepting computation. Easily, M' can be constructed out of one copy of M and two new states.

Similarly, if M is an m-state 1NFA for $\overline{L}$, then a 1NFA M' for $\overline{L^{\#}}$ can simply simulate M on a nondeterministically chosen #-delimited infix of its input, and accept if the simulation accepts; at the same time, additional nondeterministic threads accept if the input fails to be a #-delimited concatenation of infixes. Easily, M' can be constructed out of one copy of M and four new states.   □

Two upper bounds for Π'_n. It is now enough to prove that each of Π'_n and $\overline{\Pi'_n}$ can be solved by a 1NFA with $O(n^2)$ states. To see how, let us first suppose that the input is promised to be of the form $\Gamma_n^* X_n \Delta_n^*$.

It is easy to see that such an input is in Π'_n iff it contains two disjoint paths that run from the leftmost to the rightmost column and have their right endpoints in the same order as their left endpoints. *To verify this condition,* a 1NFA M can simply guess the two paths (at each step remembering only the most recent node in each of them) and accept iff their last nodes are in the order in which the paths started. This can be done easily with $2\binom{n}{2}$ states. *To disprove this condition,* a 1NFA $\overline{M}$ can look for one of the following 'flaws': (i) in some $a \in \Gamma_n$, one of the roots touches two roots of the following symbol, (ii) in some $a \in \Delta_n$, one of the roots touches two roots of the preceding symbol, or (iii) the input (is well-formed, but) does not respect the tree order. The last flaw can be detected easily, with a slightly modified copy of M; detecting (ii) is then possible with one additional state; a final modification—requiring $\binom{n}{2}$ new states—ensures that (i) is also detected. Overall, $1 + 3\binom{n}{2}$ states are enough.

Now, if the input is not promised to be of the form $\Gamma_n^* X_n \Delta_n^*$, we can simply augment M and $\overline{M}$ to also check this additional condition. Specifically, given that $\Gamma_n^* X_n \Delta_n^*$ can be recognized by a 1DFA M' with only two states, Π'_n can be solved by the (standard) Cartesian product of M and M' that accepts iff both of them accept (and is twice as big as M); similarly, $\overline{\Pi'_n}$ can be solved by an augmented version of $\overline{M}$ that includes M' as an additional nondeterministic thread (and has two more states than $\overline{M}$).

4   The Lower Bound

Much like what we did in Section 3, we first reduce the task of proving a lower bound for SDFAs solving Π_n to the task of proving a lower bound for a simpler class of automata (the *parallel intersection automata,* see below) solving Π'_n. Essential in this reduction is the notion of *generic strings* (adapted from [6]). So, we start with the definition and properties of these strings, continue with the reduction, and conclude with the lower bound for the simpler setting.

Generic strings. Let M be a SDFA over an alphabet Σ and state set Q. For any $y \in \Sigma^*$, consider the set of all states that can be produced on the rightmost boundary of y by left computations of M:

$$\text{LVIEWS}_M(y) := \{q \in Q \mid (\exists p \in Q)[\text{LCOMP}_{M,p}(y) \text{ exits into } q]\}.$$

How does this set change if we replace y with some right-extension yz of it? In other words, how do the sets $\text{LVIEWS}_M(y)$ and $\text{LVIEWS}_M(yz)$ compare?

Consider the partial function $\text{LMAP}_M(y,z) : \text{LVIEWS}_M(y) \to Q$ which, for every $q \in \text{LVIEWS}_M(y)$, is defined only if $\text{LCOMP}_{M,q}(z)$ does not hang and, if so, returns the state that this computation exits into. Easily, the values of this function: (i) are all in $\text{LVIEWS}_M(yz)$, and (ii) cover the entire $\text{LVIEWS}_M(yz)$.[1] So, $\text{LMAP}_M(y,z)$ is a *partial surjection* from $\text{LVIEWS}_M(y)$ to $\text{LVIEWS}_M(yz)$. This immediately implies Fact 1. Fact 2 is equally simple.

Fact 1. *For all y, z: $|\text{LVIEWS}_M(y)| \geq |\text{LVIEWS}_M(yz)|$.*

Fact 2. *For all y, z: $\text{LVIEWS}_M(yz) \subseteq \text{LVIEWS}_M(z)$.*

Now consider any property $\emptyset \neq P \subseteq \Sigma^*$ which is *infinitely extensible to the right*, in the sense that every string that has the property can be right-extended into a longer one that also has it. Fact 1 implies the following about the behavior of M on P: if we start with any $y \in P$ and keep right-extending it ad infinitum into $yz, yzz', yzz'z'', \cdots \in P$, then from some point on the corresponding sequence of the sizes of the sets $|\text{LVIEWS}_M(\cdot)|$ will become constant. Any of the extensions after that point is called L-generic (for M) over P. Summarizing:

Definition 1. *A string y is* L-generic *over P if $y \in P$ and, for all $yz \in P$, $|\text{LVIEWS}_M(y)| = |\text{LVIEWS}_M(yz)|$.*

Fact 3. *Suppose $P \subseteq \Sigma^*$ is non-empty and infinitely extensible to the right. Then* L-*generic strings over P exist.*

Note that a symmetric argument works in the other direction, too: working with right computations and left-extensions, we can define $\text{RVIEWS}_M(y)$ and $\text{RMAP}_M(z,y)$; conclude Facts 1 and 2 for $\text{RVIEWS}_M(y)$ and $\text{RVIEWS}_M(zy)$; define R-generic strings; and conclude Fact 3 for them, too. In fact, we can often construct strings, called simply *generic*, that are simultaneously L- and R-generic:

Fact 4. *Suppose that y_L and y_R are* L-*generic and* R-*generic over P, respectively. Then every string in P of the form $y_\text{L} z y_\text{R}$ is generic over P.*

Proof. For any L-generic string over P, all right-extensions of it in P are clearly also L-generic. In the other direction, the symmetric statement is true.   □

The next lemma is the key for the reduction presented in Lemma 4.

Lemma 3. *Suppose a SDFA M solves $L^\#$ and y is generic for it over $L^\#$. Then a string x belongs to L iff $\text{LMAP}_M(y, xy)$ and $\text{RMAP}_M(yx, y)$ are total and injective.*

Proof. Suppose $x \in L$. Since $y \in L^\#$ (because y is generic over $L^\#$), we know yxy is also in $L^\#$. Hence, yxy is a right-extension of y in $L^\#$. Since y is L-generic, this implies that $|\text{LVIEWS}_M(y)| = |\text{LVIEWS}_M(yxy)|$.

Now consider $\text{LMAP}_M(y, xy)$. By the discussion before Fact 1, we already know this is a partial surjection from $\text{LVIEWS}_M(y)$ to $\text{LVIEWS}_M(yxy)$. Since the two sets are of equal size, the function must be total. For the same reason, it must also be injective. The argument for $\text{RMAP}_M(yx, y)$ is symmetric.

Conversely, suppose $\text{LMAP}_M(y, xy)$ is total and injective. Since we already know that it partially surjects $\text{LVIEWS}_M(y)$ to $\text{LVIEWS}_M(yxy)$, we can conclude that it is actually a bijection between the two sets. Now, by Fact 2, we also know that $\text{LVIEWS}_M(yxy) \subseteq \text{LVIEWS}_M(y)$. Hence, $\text{LMAP}_M(y, xy)$ bijects $\text{LVIEWS}_M(y)$ into one of its subsets. Clearly, this is possible only if this subset is the set itself. So, $\text{LMAP}(y, xy)$ is a permutation π of $\text{LVIEWS}_M(y)$. Symmetrically, if $\text{RMAP}_M(yx, y)$ is total and injective, then it is a permutation ρ of $\text{RVIEWS}_M(y)$.

Now pick any $k \geq 1$ such that each of π^k and ρ^k is the identity on its domain, and consider the string $z := y(xy)^k = (yx)^k y$. It is easy to verify that $\text{LMAP}_M(y, (xy)^k)$ equals $\text{LMAP}_M(y, xy)^k = \pi^k$, and is therefore the identity on $\text{LVIEWS}_M(y)$. Similarly, $\text{RMAP}_M((yx)^k, y)$ equals ρ^k, and is therefore the identity on $\text{RVIEWS}_M(y)$. Intuitively, this means that, computing through z, the left-to-right computations of M do not notice the presence of $(xy)^k$ to the right of the prefix y; similarly, the right-to-left computations do not notice the presence of $(yx)^k$ to the left of the suffix y. Consequently, M does not distinguish between y and z: it either accepts both of them or rejects both of them. Since M solves $L^\#$ and $y \in L^\#$, we know M accepts y. Therefore, M accepts z as well. Hence, every #-delimited infix of z is in L. In particular, $x \in L$. $\qquad\square$

Parallel intersection automata. A *parallel intersection automaton* over Σ is any pair $M = (\mathcal{L}, \mathcal{R})$ of families of 1DFAs over Σ. To run M on an input x means to run each of its component 1DFAs on x, but with a twist: each $D \in \mathcal{L}$ reads x from left to right, while each $D \in \mathcal{R}$ reads x from right to left. We say M *accepts* x iff all these computations are accepting—i.e., iff all $D \in \mathcal{L}$ accept x and all $D \in \mathcal{R}$ accept x^R. The next lemma presents a non-trivial connection with SDFAs—implicitly present already in the argument of [6].

Lemma 4. *If $L^\#$ can be solved by a SDFA of size m, then L can be solved by a parallel intersection automaton with at most $2\binom{m}{2}$ components, each of size $\binom{m}{2}$.*

Proof. Suppose a SDFA M over a set Q of m states solves $L^\#$. We will construct a parallel intersection automaton $M' = (\mathcal{L}, \mathcal{R})$ that solves L, as follows.

First, we fix y to be any generic string for M over $L^\#$ (we know such y exist, by Facts 3,4 and easy properties of $L^\#$). Then (Lemma 3) an arbitrary x is in L iff $\text{LMAP}_M(y, xy)$ and $\text{RMAP}_M(yx, y)$ are both total and injective, namely iff:

- for all distinct $p, q \in \text{LVIEWS}_M(y)$: both $\text{LCOMP}_{M,p}(xy)$ and $\text{LCOMP}_{M,q}(xy)$ exit xy, and they do so into different states, and
- for all distinct $p, q \in \text{RVIEWS}_M(y)$: both $\text{RCOMP}_{M,p}(yx)$ and $\text{RCOMP}_{M,q}(yx)$ exit yx, and they do so into different states.

Letting $m_{\text{L}} := |\text{LVIEWS}_M(y)|$ and $m_{\text{R}} := |\text{RVIEWS}_M(y)|$, we see that checking $x \in L$ reduces to checking $\binom{m_{\text{L}}}{2} + \binom{m_{\text{R}}}{2}$ separate conditions, one for each unordered pair of distinct states from $\text{LVIEWS}_M(y)$ or from $\text{RVIEWS}_M(y)$. The components of M' are designed to check exactly these conditions.

Before describing these components, let us rewrite the above conditions a bit more nicely. First, we need a concise way of saying whether two left computations on y exit into different states or not, and similarly for right computations. To this end, we define the following relations on Q:

- $p \asymp_{\text{L}} q$ iff both $\text{LCOMP}_{M,p}(y)$ and $\text{LCOMP}_{M,q}(y)$ exit y, and they do so into different states.
- $p \asymp_{\text{R}} q$ iff both $\text{RCOMP}_{M,p}(y)$ and $\text{RCOMP}_{M,q}(y)$ exit y, and they do so into different states.

Now, the conditions from above can be rephrased as follows:

- for all distinct $p, q \in \text{LVIEWS}_M(y)$: both $\text{LCOMP}_{M,p}(x)$ and $\text{LCOMP}_{M,q}(x)$ exit x, and they do so into states that are $\asymp_{\text{L}}$-related, and
- for all distinct $p, q \in \text{RVIEWS}_M(y)$: both $\text{RCOMP}_{M,p}(x)$ and $\text{RCOMP}_{M,q}(x)$ exit x, and they do so into states that are $\asymp_{\text{R}}$-related,

and it is now straightforward to build 1DFAs that check each of them.

For example, the 1DFA checking the condition for the pair $p, q \in \text{LVIEWS}_M(y)$ has 1 state for each unordered pair of distinct states from Q, with $\{p, q\}$ being both the start and the accept state. On $\vdash$, $\{p, q\}$ simply goes to itself. At every step after that, the automaton tries to compute the next pair by applying the transition function of M on the current symbol and each of the two states of the current pair. If either application returns no value or both return the same value, the automaton simply hangs; else, it moves to the corresponding pair. On $\dashv$, the pairs leading to $\{p, q\}$ (and thus to acceptance) are exactly the $\asymp_{\text{L}}$-related ones.

Overall, we need $\binom{m_{\text{L}}}{2} + \binom{m_{\text{R}}}{2} \leq 2\binom{m}{2}$ automata, each of size $\binom{m}{2}$. $\qquad\square$

A lower bound for Π'_n. By Lemma 4, it is now enough to prove that no parallel intersection automaton can solve Π'_n with a *small number* of *small components*. The next lemma proves something much stronger: no parallel intersection automaton can solve Π'_n with small components, irrespective of their number. The argument is similar to that of [5, Theorem 4.2.3].

Lemma 5. *In any parallel intersection automaton solving Π'_n, at least one of the components has size strictly greater than $(2^n - 2)/n$.*

Proof. Towards a contradiction, suppose $M = (\mathcal{L}, \mathcal{R})$ solves Π'_n with at most $(2^n - 2)/n$ states in each one of its components. We can then prove the following.

Claim. There exists a string $u \in \Gamma_n^*$ that admits well-formed right-extensions and has all of them accepted by every $D \in \mathcal{L}$. Symmetrically, some $v \in \Delta_n^*$ admits well-formed left-extensions and has all of them accepted by every $D \in \mathcal{R}$.

Intuitively, u is a string that manages to 'confuse' every left component of M: each of them accepts every well-formed right-extension of u (no matter whether

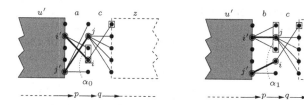

Fig. 2. Confusing D in the proof of Lemma 5

it respects the tree order or not), exactly because it has failed to correctly keep track of the tree order inside u. Similarly for v and the right components of M.

We will prove only the first half of the claim, as the argument for the other half is symmetric. Before that, though, let us see how the claim implies a contradiction. First, since u has well-formed right-extensions, we can find nodes $i, j \in [n]$ on its rightmost column that belong to different trees. Similarly, the leftmost column of v contains nodes $r, s \in [n]$ that belong to different trees of v. Now, consider the two symbols of X_n that have i, j, r, s as their roots, namely $x := \{(i, r), (j, s)\}$ and $x' := \{(i, s), (j, r)\}$, and the strings uxv and $ux'v$. Clearly, each string is well-formed, right-extends u, and left-extends v. So, by the claim, each of them is accepted by all components of M. Hence, M accepts both strings. However, by the selection of x and x', we know that one of the strings does not respect the tree order. So, after all, M does not solve Π'_n—a contradiction.

To prove the first half of the claim, we work by induction on the size of $\mathcal{L}$.

If $\mathcal{L}$ is empty, then the claim holds vacuously for, say, the empty u.

If $\mathcal{L}$ is non-empty, we pick any D in it and let $\mathcal{L}' := \mathcal{L} - \{D\}$. Then $\mathcal{L}'$ is smaller than $\mathcal{L}$, so (by the inductive hypothesis) some $u' \in \Gamma_n^*$ admits well-formed right-extensions and has all of them accepted by all $D' \in \mathcal{L}'$. Our goal is to find two symbols $a, c \in \Gamma_n$ such that the string $u := u'ac$ admits well-formed right-extensions and has all of them accepted by all members of $\mathcal{L}$. (Fig. 2.)

We start by noting (as above) that, since u' has well-formed right-extensions, there exist nodes i' and j' in its rightmost column that belong to different trees.

Moreover, some of the well-formed right-extensions of u' respect the tree order (because, for each extension that does not, there is one that does: the one that differs only in the pairing of the roots of the middle symbol) and are therefore accepted by M. In particular, they are accepted by D. Thus, the left computation of D on each of them exits to the right. Hence, the left computation of D on u' exits to the right, too. Let p be the corresponding exit state.

Based on D, i', j', and p, we can now find the symbols a, c that we are after.

Consider all symbols of Γ_n that have i' and j' as roots. Each of them is of the form (i', j', α) and takes p to some next state. Since there are $2^n - 2$ such symbols (one for each $\emptyset \neq \alpha \subsetneq [n]$) and D has at most $(2^n - 2)/n$ states, we know some next state attracts at least $(2^n - 2)/((2^n - 2)/n) = n$ symbols. Call this state q. Among the α's that correspond to the symbols taking p to q, two must be incomparable (otherwise, they would form a chain of n or more non-trivial subsets of $[n]$—a contradiction). Call these subsets α_0 and α_1. Then symbol a is one of the two corresponding symbols, say $a := (i', j', \alpha_0)$. We also name the other symbol, say $b := (i', j', \alpha_1)$, and a node in each side of the symmetric difference of the two sets, say $i \in \alpha_0 \setminus \alpha_1$ and $j \in \alpha_1 \setminus \alpha_0$ (both sides are non-empty, by the incomparability of α_0, α_1). It is important to note that a connects i' and j' to i and j, respectively, whereas in b this connection is reversed. Finally, c is selected to be any symbol with i and j as roots, say $c := (i, j, \{1\})$.

Let us see why $u = u'ac$ is the string that we want (ubc would also do).

First, by the choice of i' and j', we know that a extends both trees of u': one to α_0, the other one to $\overline{\alpha_0}$. Similarly, c extends both trees of $u'a$, since $i \in \alpha_0$ and $j \in \overline{\alpha_0}$. Hence, $u = u'ac$ can indeed be right-extended into well-formed strings.

Second, every such extension of u is obviously a well-formed right-extension of u', and is thus accepted by all $D' \in \mathcal{L}'$ (recall the inductive hypothesis).

Finally, every such extension of u, say uz, is also accepted by D. To see why, consider the computations of D on $u'a$ and $u'b$. Both exit into q (by the selection of a, b, q). So, the computation of D on $uz = u'acz$ has the same suffix as the computation of D on $u'bcz$. Hence, D either accepts both strings or rejects both strings. In the latter case, M would also reject both strings, contradicting the fact that one of them respects the tree order (the strings differ only at a and b, which connect i' and j' to i and j differently). Hence, D must be accepting both strings. In particular, it accepts $u'acz = uz$.   □

5   A Bigger Picture

Our theorem is only a piece in the puzzle defined by the study of size complexity in finite automata. An elegant theoretical framework for describing this puzzle is due to Sakoda and Sipser [5]. Analogous to the framework built on other computational models and resources (e.g., Turing machines and time), it is based on the notions of a *reduction* and of a *complexity class*. However, a member of a class in this framework is always a *family of problems* and each class contains exactly every family that is solvable by a family of small automata of a corresponding type. For example, 1D *contains exactly every family of problems that can be solved by some family of small* 1DFAs. Similarly, the classes 1N, 2D, and 2N were defined for 1NFAs, 2DFAs, and 2NFAs, respectively, while co1D, co1N, co2D, and co2N were defined to consist of the corresponding families of complements.

Replacing 1DFAs with SDFAs, SP₀FAs, or SNFAs in the above definition, we can naturally define the classes SD, SP₀X, and SN, respectively, for sweeping and/or LasVegas automata.[1] Then, SD $\subseteq$ SP₀X $\subseteq$ SN (trivially), $\Pi \in$ 1N $\cap$ co1N $\subseteq$ SP₀X (by Sect. 3), $\Pi \notin$ SD (by Sect. 4), and therefore SD $\not\supseteq$ SP₀X (our theorem; note that we have actually proved a stronger fact: SD $\not\supseteq$ 1N $\cap$ co1N). At the same time, we also have SP₀X $\subseteq$ SN $\cap$ coSN (trivially) and coSN $\not\supseteq$ SN (by [3]), so that SP₀X $\not\supseteq$ SN. Overall, the trivial chain SD $\subseteq$ SP₀X $\subseteq$ SN is actually SD $\subsetneq$ SP₀X $\subsetneq$ SN.

---

[1] Note the "X" in "SP₀X". The name "SP₀" is reserved for the more natural class where the SP₀FAs must run in *polynomial* expected time. Similarly for 2P₀X, RP₀X, SP₁X, etc.

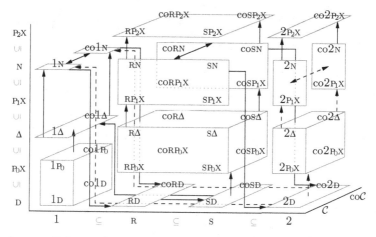

Fig. 3. A map of classes: boxes mean equality; the axes show the easy inclusions; a solid arrow $A \to B$ means $A \not\supseteq B$; a dashed arrow $A \to B$ means we conjecture $A \not\supseteq B$

Figure 3 shows in more detail the relations between the several classes, including those for *Monte-Carlo* automata ("P_1" and "P_2"—for one-sided and two-sided error), *self-verifying* automata ("Δ"—these capture the intersection of nondeterminism and co-nondeterminism; e.g., $1\Delta = 1N \cap co1N$), and *rotating* automata ("R"—these are sweeping automata capable of only left-to-right sweeps).

Most facts on this map are trivial, or easy, or modifications/consequences of known results [1–6] and of our main theorem. Exceptions include the ability of small nondeterministic and probabilistic rotating automata to simulate their sweeping counterparts: $RN = SN$, $RP_0X = SP_0X$, $RP_1X = SP_1X$, and $RP_2X = SP_2X$. A more detailed presentation will appear in the full version of this article.

Acknowledgement. We would like to thank Juraj Hromkovič for suggesting the problem and for giving us many helpful comments.

References

1. Hromkovič, J., Schnitger, G.: On the power of LasVegas for one-way communication complexity, OBDDs, and finite automata. Information and Computation 169, 284–296 (2001)
2. Hromkovič, J., Schnitger, G.: On the power of LasVegas II: two-way finite automata. Theoretical Computer Science 262(1–2), 1–24 (2001)
3. Kapoutsis, C.A.: Small sweeping 2NFAs are not closed under complement. In: Bugliesi, M., Preneel, B., Sassone, V., Wegener, I. (eds.) ICALP 2006. LNCS, vol. 4051, pp. 144–156. Springer, Heidelberg (2006)
4. Macarie, I.I., Seiferas, J.I.: Strong equivalence of nondeterministic and randomized space-bounded computations. Manuscript (1997)
5. Sakoda, W.J., Sipser, M.: Nondeterminism and the size of two way finite automata. In: Proceedings of the STOC, pp. 275–286 (1978)
6. Sipser, M.: Lower bounds on the size of sweeping automata. Journal of Computer and System Sciences 21(2), 195–202 (1980)

Stochastic Methods for Dynamic OVSF Code Assignment in 3G Networks

Mustafa Karakoc[1] and Adnan Kavak[2]

[1] Kocaeli University, Dept. of Electronics and Computer Edu., 41380, Kocaeli, Turkey
[2] Kocaeli University, Dept. of Computer Engineering, 41040, Kocaeli, Turkey
{mkarakoc,akavak}@kou.edu.tr

Abstract. Orthogonal variable spreading factor (OVSF) codes are widely used to provide variable data rates for supporting different bandwidth requirements in wideband code division multiple access (WCDMA) systems. Many works in the literature have intensively investigated to find an optimal dynamic code assignment scheme for OVSF codes. Unlike earlier studies, which assign OVSF codes using conventional (CCA) or dynamic (DCA) code allocation schemes, in this paper, stochastic optimization methods which are genetic algorithm (GA) and simulated annealing (SA) were applied which population is adaptively constructed according to existing traffic density in the OVSF code-tree. Also, the influences of the GA (selection, crossover and mutation techniques) and the SA (cooling schedules, number of inner loop) parameters were examined on the dynamic OVSF code allocation problem. Simulation results show that the GA and SA provide reduced code blocking probability and improved spectral efficiency when compared to the CCA and DCA schemes.

1 Introduction

In order to meet the demands of mixed traffic applications, 3G systems support variable data rates for different users. WCDMA is the most popular 3G radio access technology. In WCDMA systems [1], orthogonal variable spreading factor (OVSF) codes are used to facilitate variable rate data transmissions. Each base station (BS) in WCDMA manages a code tree for downlink transmission. Since the OVSF codes are limited, effective management of this resource is an important issue. There are several techniques, which are mainly classified as the conventional code allocation (CCA) [2] and the dynamic code allocation (DCA) [3] schemes. CCA scheme basically assigns an OVSF code for an incoming call request if there is an available one in the code tree, otherwise call is dropped. DCA schemes reallocate some used codes in the code tree in order to find a suitable code for the call request. In this paper, we focused on stochastic search techniques such as Genetic Algorithm (GA) and Simulated Annealing (SA) for dynamic allocation of OVSF code tree with a random initial population. The remainder of this paper proceeds as follows. In the next section the background knowledge is reviewed. GA and SA based DCA schemes are presented in Section III. In Section IV simulation parameters and results. Finally, conclusions are given in Section V.

J. Hromkovič et al. (Eds.): SAGA 2007, LNCS 4665, pp. 142–153, 2007.
© Springer-Verlag Berlin Heidelberg 2007

2 Basic Background Knowledge

2.1 OVSF Code Tree

In a WCDMA system, two operations are applied to user data [4]. The first one is channelization, which transforms every bit into a code sequence. The length of the code sequence per data bit is called the spreading factor (*SF*), which is typically power of two. Channelization codes in the OVSF code tree have a unique description as $C_{SF,k}$, where k is the code number, $1 \leq k \leq SF$. The second operation is scrambling, which applies a scrambling code to the spread signal. Scrambling codes are used to separate transmission from a single source. All codes in the same layer are orthogonal to each other, while codes in different layers are also orthogonal if they do not have an ancestor-descendant relationship. The data rate is doubled whenever we go one level up in the tree.

2.2 Related Works

Many existing works have been thoroughly investigated as follows. Tseng *et al.* [5] have proposed single-code and multi-code placement and replacement schemes for WCDMA systems. The algorithm for single code placement/replacement possibly produces a code blocking problem. The multi-OVSF code placement and replacement scheme presented by Chao *et al.* [6] reduces the code blocking problem by using a code-separation operation. Minn *et al.* [3] developed a dynamic code assignment (DCA) scheme, which is based on the code pattern search to find a branch of requested rate in the code tree, which can be vacated with minimum cost. Regarding GAs [7], is generally applied in wireless communications for optimizing and designing antenna arrays [8], or detecting multiuser [9]. An OVSF code allocation strategy using GA is proposed in by Cinteza *et al* [10]. They have used binary representation of a chromosome and investigated fixed traffic density. New code requests coming onto an OVSF code tree already containing active codes are managed by using the GA. SA firstly proposed by Kirkpatrick *et al.* [11], which is based on the analogy between the process of finding best solution of a combinatorial optimization problem and annealing process. However, SA algorithm is also applied in wireless communications for detecting multiuser [12], and optimizing and designing antenna arrays [13].

3 Dynamic OVSF Code Allocation Using GA/SA

Code blocking is the major problem in OVSF code assignment, which limits system performance. This section discusses stochastic search techniques such as GA and SA in OVSF code assignment strategy. General flowchart is given at Figure 1. In idle state, execution is not required for resource assignment. Call is initiated with the call processor's signaling to resource manager to allocate resources for a traffic channel. First, availability of capacity (total rates of unused codes) is checked in the code tree whether to support the requested call rate. If there is enough capacity, then availability of requested rate OVSF code is checked among unused codes in the relevant layer,where the call can be supported. If a call cannot be assigned a code due unavailability of the code with the requested rate, GA/SA block is executed.

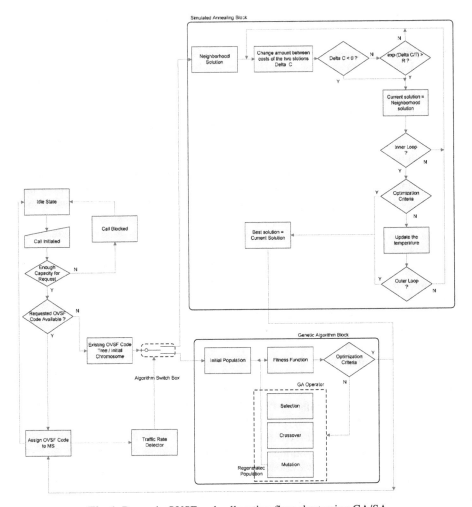

Fig. 1. Dynamic OVSF code allocation flow chart using GA/SA

This block is start with initial chromosome. For the clearly understanding of the tree structure using GA / SA, a sample OVSF code tree is shown in Figure 2. The OVSF code tree which is input to the GA / SA block is named as initial chromosome (Ch_{ini}) and this chromosome is represented with the index information belonging to active users. In other words, index numbers of occupied branches are expressed by existing OVSF code tree index which corresponds to a chromosome in the population. Traffic rate detector detects the traffic to change the algorithm in code reassignment process.

3.1   GA Based Dynamic OVSF Code Allocation Scheme

GA, one of the optimization and global search methods, is applied effectively to solve various combinatorial optimization problems and worked with probabilistic rules [7]. Selection, crossover and mutation are the most known genetic operators.

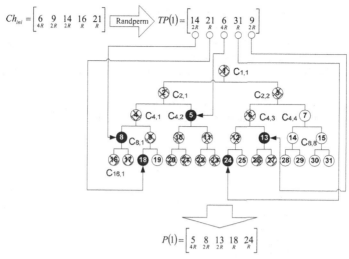

Fig. 2. Sample OVSF code tree and construction of initial population from code tree

GA starts with initial chromosome. The size of initial population P of n chromosomes is adaptively changes according to current traffic density. Each chromosome in the initial population has different code tree index number, but the number of data bit rates and the total amount data bit rates of each chromosome is identical with initial chromosome. In Figure 2, initial chromosome is Ch_{ini} =[6 9 14 16 21] and the data bit rate of this chromosome is [4R 2R 2R R R]. Each active user's index number in the chromosome is called as a gene and gene is represented by an integer number. Population size is depends on the traffic density which is

$$n = SF - \sum_{i=1}^{U} S(i) \qquad (1)$$

where, U is the total number of active users and $S(i)$ is the date rate of ith active user, where $i=1,...,U$. In the above figure, U and S are 5 and $10R$, respectively. Therefore the number of chromosome in the initial population (n) is obtained as 6 (16-10). Each of n chromosomes comprises coded information of existing OVSF code tree obtained with using random permutation that fully describes a potential solution to the optimization problem and expresses the different OVSF code tree. Nevertheless number of users and each user's data bit rate remain same as Ch_{ini}.

Figure 2 also shows how the 1st chromosome is obtained from the Ch_{ini}. Temporary population $TP(1)$ which is derived from Ch_{ini} with random permutation is sequentially assigned to empty OVSF code tree from 1st gene to Uth gene. It is important to consider the orthogonality principle, while assigning codes in the OVSF code tree. Index numbers are taken to compose a new chromosome $P(1)$. The process of obtaining $P(1)$ is as follows: For each gene of $TP(1)$, the corresponding gene in $P(1)$ is selected as the possible leftmost OVSF code that has the same rate as this gene in $TP(1)$. For instance, the first gene numbered by 14 in $TP(1)$ has the rate 2R. Hence, possible leftmost gene with rate 2R is the OVSF code numbered as 8 in $P(1)$. For, the gene numbered 21 with rate R in $TP(1)$, we obtain the OVSF code numbered 18, and so on. After obtaining each corresponding gene for $P(1)$, we list the genes in $P(1)$

from highest rate to lowest rate. This process is repeated n times to fill the P. P shows, several different possible result for a given problem. It is clear that iteration number of optimal solution is depends on population size (n), users' data bit rates ($S(i)$), and their location in the code tree.

Then, the fitness value for each chromosome of population is evaluated according to fitness function, which is defined specially for OVSF code assignment–reassignment problem. The fitness value of jth chromosome $f(j)$ is the quantity of replacement of each individual in $P(j)$ according to Ch_{ini} defined by

$$f(j) = \frac{1}{\sum_{i=1}^{U} \left[(Ch_{ini}(j) - P(j,i)) \times S(i) \right]} \tag{2}$$

where, j is the chromosome number, $j=1,...,n$. Based on the fitness values of the chromosomes in the population, the selection operator creates a new population of n chromosomes, which contains chromosomes that, on average, have better fitness values than those in the original population. In order to produce better traits, the chromosomes should be hybridized using the crossover operation. Pairs of chromosomes are selected from the population subjected to crossover rate (p_c). The number of chromosomes (n_j) with crossover operation is given by

$$n_j = n \times p_c \tag{3}$$

If n_j is odd, then $(n_j-1)/2$ randomly selected chromosome pairs are used for crossover, otherwise it is $n_j/2$. Chromosome pairs used at crossover operation are randomly chosen. Then, the mutation operation is applied to the population produced by the crossover operation to preserve genetic diversity by perturbing chromosomes randomly. In this operation, if randomly obtained number between 0 and 1 smaller than mutation rate (p_m), then mutation process is started; otherwise a new random number is taken for the next chromosome. Finally each chromosome in the population is checked in terms of its fitness values. If an OVSF code tree which represented by best chromosome, can assign the requested data bit rate to desired user, then optimization criterion is confirmed and requested data bit rate is assigned to desired user. Otherwise, another chromosomes in the population are checked. This process is run-on until to assign the requested data bit rate to new user or to met the predetermined loop. GA operators are described as follows:

Population: Because the size of the population varies according to traffic density, there is no clear mark how large it should be.

Fitness Function: This is specifically defined for OVSF code reassignment problem, which is the quantity of replacement of each individual in population according to Ch_{ini}.

Selection: In this study roulette wheel selection technique has been used. In roulette wheel, probability of each chromosome ($Pr(j)$) is inverse proportional with its fitness value ($f(j)$). This selection process continued until the population is completed.

Crossover: Crossover operator is powerful for exchanging information between chromosomes and creating new solutions. This work consider single point crossover operator in which for each pair to be crossed a random integer l is chosen as crossover

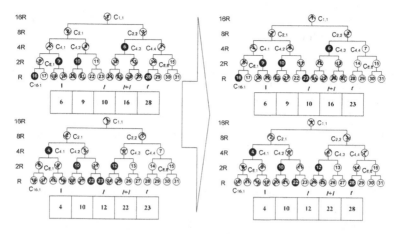

Fig. 3. Single point crossover operator

point. Randomly selected two chromosomes' first pair's head part up to *l*th gene is associated with second pair's tail part from *l+1*st gene to *l*th gene, where *t* is the length of each chromosome, as depicted in Figure 3,

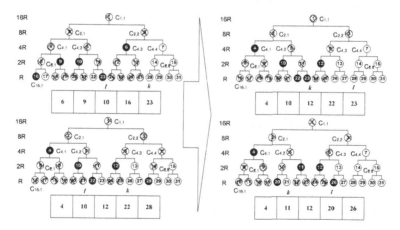

Fig. 4. Swap mutation operator

Mutation: This operator is used to prevent the reproduction of similar type chromosomes in population. Swap mutation operator used in this work, which randomly selects two genes in chromosome and swaps the positions of these genes to produce a new chromosome. Figure 4 shows swap mutation operation.

3.2 SA Based Dynamic OVSF Code Allocation Scheme

SA, firstly developed by Kirkpatrick *et al.* [11], is a local search algorithm. The searching process in SA starts with initial chromosome. A neighborhood of this solution is generated using any neighborhood move rule. Then the cost of this possible solution (=chromosome) is obtained with

$$\Delta f = f_i - f_{i-1} \tag{4}$$

where, Δf represents the change amount between costs of the two solutions. f_i and f_{i-1} represent fitness values belong to neighborhood solution and current solution, respectively. If $\Delta f < 0$, current solution is replaced with the generated neighborhood solution. Otherwise if $\Delta f > 0$, its mean that current solution is replaced with the generated neighborhood solution within the limits of specific probability.

$$\exp\left(-\Delta f / T\right) > R \tag{5}$$

where, T is temperature which is a positive control parameter and R is a random number varies from 0 to 1. Then inner loop is checked. Algorithm turns mutation operator to obtain possible solution with better fitness value, until inner loop criterion is met. Then, the optimization criterion is checked. The optimization criterion check blocks in the flow chart are used to control the algorithm. If criterion is provided, then algorithm is finalized and requested data bit rate is assigned to new user. This process is run-on until to assign the requested data bit rate to new user or to met the outer loop. If the requested data bit rate can not assign while the outer loop is met, then call is blocked. SA operators are described as follows:

Fitness Function: This is same with the GA (Eq. 2).

Neighborhood Move: This operator is used to produce a near solution to current solution in search space. Swapping move is used in this paper. This operator works same with the swap mutation operation in GA.

Cooling Schedule: The performance of this algorithm is dependant on this operator. Lundy & Mees is used in this work as a cooling schedule. In Lundy & Mees schedule; the relationship between T_{k+1} and T_k is below:

$$T_{k+1} = \frac{T_k}{1 + \beta T_k}, \ \beta = \frac{T_i - T_f}{MT_i T_f} \tag{6}$$

where, ($\beta > 0$) is coefficient between two temperatures, T_{k+1} and T_k. In this study, initial temperature (T_i) is 0.9, while final temperature (T_f) is 0.1, and M is the number of outer loop in the algorithm.

Inner Loop and Outer Loop: Inner loop criterion decides how many possible new solution produced in every temperature. Outer loop criterion is used to stop the searching process. In this study, inner loop criterion is set to 5. Outer loop criterion is number of 1000.

4   Performance Evaluation

For performance measurement, firstly call blocking probability is investigated, which is the ratio of the number of blocked calls to the number of incoming call requests. Then, spectral efficiency is calculated as the ratio of data rates of served calls to the total requested data rate. Finally influences of several GA and SA operators are considered. In the simulations: Representation of chromosomes is integer. Population size is depending on the density of traffic. In GA sub-block roulette wheel selection, single point crossover and swap mutation techniques are used. Mutation rate (p_m) and crossover rate (p_c) are taken as 0.2 and 0.8, respectively. In SA sub-block initial and final temperatures are 0.9 and 0.1, respectively. Swap neighborhood move and Lundy & Mees cooling schedule are used. Inner loop criteria is set to 5, while outer loop is 1000. However, call arrival process is Poisson with mean arrival rate of λ varied from 4 to 64 calls/unit. Call duration is exponentially distributed with a mean value of μ is 0.25 units of time. Spreading factor (*SF*) is 256. Possible OVSF code rates are generated using uniform distribution between R and *SF*×R. Single simulation is performed until at least 1000 incoming calls. For the same input parameters, the simulations are repeated 10 times, and then results are averaged.

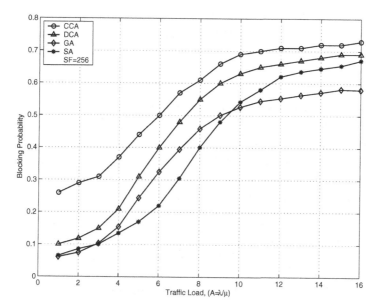

Fig. 5. Blocking probability at different traffic loads, *SF*: 256

Figure 5 shows the simulation results for blocking probability. It is clearly seen that GA and SA have less blocking probability than DCA and CCA. When we compare the GA and SA, we can se that the blocking probability of GA and SA are depend on the traffic load. For example, when the traffic load is 6, blocking probability of GA, SA, DCA, and CCA are % 22, % 32.5, % 40, % 50, respectively. If the traffic load is increased to 12, blocking probability of GA, SA, DCA, and CCA are % 62.1, % 55.3, % 65.9, % 70.9, respectively. As a result, when the traffic load is increased from 6 to 12, the algorithm with best performance shift from SA to GA.

Figure 6 shows the spectral efficiency of the GA, SA, DCA, and CCA methods at different traffic loads. The spectral efficiency of the resource is inversely proportional to the traffic load in the system. It is clearly seen that GA and SA are effectively use given spectrum. As numerically at traffic load 10, spectral efficiency of GA, SA, DCA, and CCA are % 12.6, % 22.2, % 25.6, % 25.9, respectively.

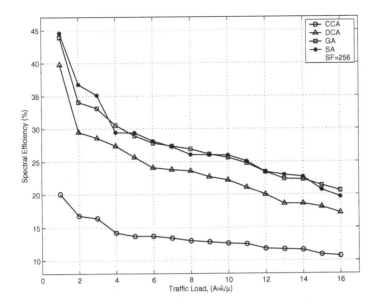

Fig. 6. Spectral efficiency at different traffic loads, *SF*: 256

Also, the influences of the GA and SA parameters were examined on the dynamic OVSF code allocation problem. In order to improve the system performance, different selection, crossover, and mutation techniques are investigated. For this purpose, roulette wheel and tournament methods as selection, single point and arithmetic methods as crossover, and swap and shift methods as mutation, are compared. In selection, the tournament operator uses the roulette selection N times to produce a tournament subset of chromosomes, where the tournament size N is chosen as random integer number. The best chromosome is then chosen as the selected chromosome. However, arithmetic crossover technique linearly combines parent chromosomes to produce new chromosomes according to the following equations:

$$\text{Offspring\_1} = a * \text{Parent\_1} + (1-a) * \text{Parent\_2}$$
$$\text{Offspring\_2} = (1-a) * \text{Parent\_1} + a * \text{Parent\_2}$$

$$(7)$$

where, a is a random weighting factor. As close as rounded integer values are taken for offspring_1 and offspring_2. In shift mutation technique, a selected gene in the chromosome is shifted either left hand or right hand side. Table 1 considers the results for the average number of reassigned users with several selection, crossover, and mutation techniques.

According to Table 1, it is clearly seen that, performance improvements are proportional with traffic load. In traffic load 10, number of reassigned user is more at tournament method (18.01) than roulette wheel method (11.72) for selection, more at arithmetic method (15.68) than single point method (11.72) for crossover, more at shift method (13.98) than swap method (11.72) for mutation techniques.

Table 1. Influences of selection, crossover, and mutation operators

	Traffic Density (λ)							
	8	**16**	**24**	**32**	**40**	**48**	**56**	**64**
Selection								
Roulette Wheel	4.51	5.49	6.72	8.86	11.72	16.68	20.88	26.02
Tournament	6.92	8.91	11.41	14.74	18.01	21.72	30.06	42.17
Crossover								
Single Point	4.51	5.49	6.72	8.86	11.72	16.68	20.88	26.02
Arithmetic	5.29	8.12	9.93	11.34	15.68	20.14	22.4	24.77
Mutation								
Swap	4.51	5.49	6.72	8.86	11.72	16.68	20.88	26.02
Shift	4.74	5.83	6.34	8.04	13.98	20.47	26.18	31.62

Table 2. Average number of reassigned users with varying cooling schedules

	Traffic Density (λ)							
	8	**16**	**24**	**32**	**40**	**48**	**56**	**64**
Cooling Schedules								
Lundy & Mees	4.51	5.49	6.72	8.86	11.72	16.68	20.88	26.02
Proportional	3.19	3.6	4.54	5.17	7.23	10.42	16.0	19.78

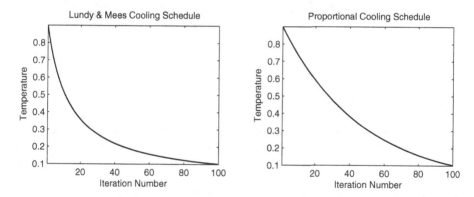

Fig. 7. Lundy & Mees and proportional cooling schedules

Selection, crossover, and mutation techniques are related to GA sub-block. In order to investigate the effect of SA sub-block, several cooling schedules and different number of inner loops are considered. Step size of algorithm in the search space is defined with cooling schedule. Here, Lundy & Mees cooling schedule is compared with proportional to calculate T_{k+1} from T_k. In proportional cooling schedule,

$$T_{k+1} = \alpha T_k, \quad \alpha = \sqrt[M]{\frac{T_f}{T_i}} \tag{8}$$

Table 2 presents the comparison of Lundy & Mees and proportional cooling schedules. The more number of reassigned users is seen when Lundy & Mees cooling schedule is used. It is also clear that the number of reassigned user is increased with traffic rate.

Figure 7 shows the Lundy & Mees and proportional cooling schedule. Because of the Lundy & Mees converges faster than proportional to lower temperature values for each simulation step, this schedule gives better performance.

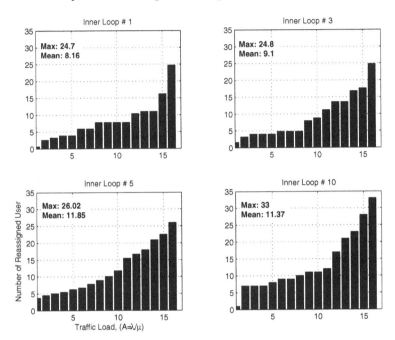

Fig. 8. Number of reassigned users with different number of inner loops

The number of reassigned users is changed using different number of inner loops. Inner loop in the ASAGA decides, how many new solutions are generated for each temperature value. Possible solution with best fitness value is obtained among the new solutions. For this purpose 1, 3, 5, and 10 numbers of inner loop are considered to observe the system performance. Figure 8 shows the number of reassigned users for each traffic load for different inner loop numbers. It is clear that the more number of new solution means the better fitness value obtained for each temperature value. Therefore the number of inner loop is proportional with the number of reassigned users. We can also say that this proportion is valid between the number of reassigned users with traffic load.

5 Conclusions

This work mainly investigates Genetic Algorithm (GA) and Simulated Annealing (SA) based OVSF code assignment in order to provide the answer to following questions; what can be done in order to improve the number of active users in the system and how the high data bit rate requests are assigned if enough capacity is provided by system. For this purpose, the performance of GA is evaluated under different selection crossover, and mutation techniques, and SA is evaluated under different cooling schedules, and number of inner loops. The simulation results show that the GA and SA as compared to the CCA and DCA, provides the smallest blocking probability and largest spectral efficiency in the system. In addition to that different selection, crossover, and mutation operations are considered. Tournament in the selection part, arithmetic in the crossover part, and shift in the mutation part provide better performance than other techniques in each class for GA sub-block. In SA sub-block, influences of different cooling schedules and different number of inner loops are considered. It is clearly seen that Lundy & Mees has shown better performance for cooling schedule. Also, the number of reassigned users is proportional with number of inner loop and traffic load.

References

[1] Holma, H., Toksala, A.: WCDMA for UMTS. Wiley, Nokia, Finland (2000)
[2] Adachi, F., Sawahashi, M., Okawa, K.: Tree structured generation of orthogonal spreading codes with different lengths for forward link of DS-CDMA mobile. IEE Electronics Letters 33, 27–28 (1997)
[3] Minn, T., Siu, K.Y.: Dynamic Assignment of Orthogonal Variable- Spreading-Factor Codes in W-CDMA. IEEE Journal on Selected Areas in Communications 18(8), 1429–1440 (2000)
[4] Okawa, K., Adachi, F.: Orthogonal Forward Link Using Orthogonal Multi Spreading Factor Codes for Coherent DS-CDMA Mobile Radio. IEICE Transactions on Communication E81-B(4), 778–779 (1998)
[5] Tseng, Y.C., Chao, C.M.: Code Placement and Replacement Strategies for Wideband CDMA OVSF Code Tree Management. IEEE Transactions on Mobile Computing 1(4), 293–302 (2002)
[6] Chao, C.M., Tseng, Y.C., Wang, L.C.: Reducing Internal and External Fragmentations of OVSF Codes in WCDMA Systems with Multiple Codes. In Proc. IEEE Wireless Communications and Networking Conf. 1, 693–698 (2003)
[7] Goldberg, D.E.: Genetic Algorithms in Search, Optimization, and Machine Learning. Addison-Wesley, Reading, MA (1989)
[8] Manetta, L., Ollino, L., Schillaci, M.: Use of an Evolutionary Tool for Antenna Array Synthesis. In: Rothlauf, F., Branke, J., Cagnoni, S., Corne, D.W., Drechsler, R., Jin, Y., Machado, P., Marchiori, E., Romero, J., Smith, G.D., Squillero, G. (eds.) Applications of Evolutionary Computing. LNCS, vol. 3449, pp. 245–253. Springer, Heidelberg (2005)
[9] Shayesteh, M.G., Menhaj, M.B., Nobary, B.G.: A New Modified Genetic Algorithm for Multiuser Detection in DS/CDMA Systems. In: Reusch, B. (ed.) Computational Intelligence. Theory and Applications. LNCS, vol. 2206, p. 608. Springer, Heidelberg (2001)
[10] Cinteza, M., Radulescu, T., Marghescu, I.: Orthogonal Variable Spreading Factor Code Allocation Strategy Using Genetic Algorithms. Cost 290, Technical Document, Traffic and QoS Management in Wireless Multimedia Networks (Wi-QoST) (2005)
[11] Kirkpatrick, S., Gelatt, J.C.D., Vecchi, M.P.: Optimization by Simulated Annealing. Science 220, 671–680 (1983)
[12] Yoon, S.H., Rao, S.S.: Annealed neural network based multiuser detector in code division multiple access communications. IEE Communications Proceedings 147(1), 57–62 (2000)
[13] Osseiran, A., Logothetis, A.: A method for designing fixed multibeam antenna arrays in WCDMA systems. Antennas and Wireless Propagation Letters 5, 41–44 (2006)

On the Support Size of
Stable Strategies in Random Games⋆

Spyros C. Kontogiannis[1,2] and Paul G. Spirakis[2]

[1] Department of Computer Science, University Campus, 45110 Ioannina, Greece
kontog@cs.uoi.gr
[2] R.A. Computer Technology Institute, N. Kazantzakis Str., University Campus,
26500 Rio–Patra, Greece
{kontog,spirakis}@cti.gr

Abstract. In this paper we study the support sizes of evolutionary stable strategies (ESS) in random evolutionary games. We prove that, when the elements of the payoff matrix behave either as uniform, or normally distributed independent random variables, almost all ESS have support sizes $o(n)$, where n is the number of possible types for a player. Our arguments are based exclusively on the severity of a *stability* property that the payoff submatrix indicated by the support of an ESS must satisfy. We then combine our normal–random result with a recent result of McLennan and Berg (2005), concerning the expected number of Nash Equilibria in normal–random bimatrix games, to show that the expected number of ESS is significantly smaller than the expected number of symmetric Nash equilibria of the underlying symmetric bimatrix game.

JEL Classification Code: C7 – Game Theory and Bargaining Theory.

Keywords: Bimatrix Games, Evolutionary Games, Evolutionary Stable Strategies, Nash Equilibria.

1 Introduction

In this work we study the distribution of the support sizes of evolutionary stable strategies (ESS) in random evolutionary games, whose payoff matrices have elements that behave as independent, identically distributed random variables. Arguing about the existence of a property in random games may actually reveal information about the (in)validity of the property in the vast majority of payoff matrices. In particular, a vanishing probability of ESS existence would prove that this notion of stability is rather rare among payoff matrices, dictating the need for a new, more widely applicable notion of stability. Etessami and Lochbihler [2] recently proved both the **NP**−hardness and **coNP**−hardness of even detecting the existence of an ESS for an arbitrary evolutionary game.

⋆ This work was partially supported by the 6th Framework Programme under contract 001907 (DELIS).

J. Hromkovič et al. (Eds.): SAGA 2007, LNCS 4665, pp. 154–165, 2007.
ⓒ Springer-Verlag Berlin Heidelberg 2007

The concept of ESS was formally introduced by Maynard–Smith and Price [14]. Haigh [4] provided an alternative characterization, via a set of necessary and sufficient conditions (called *feasibility, superiority* and *stability* conditions), for a strategy $\mathbf{x}$ being an ESS. As will be clear later, the first two conditions imply that the profile $(\mathbf{x}, \mathbf{x})$ is a symmetric Nash equilibrium (SNE) of the underlying symmetric bimatrix game $\langle A, A^T \rangle$.

A series of works about thirty years ago (eg, [3], [9], [6]) have investigated the probability that an evolutionary game with an $n \times n$ payoff A whose elements behave as uniform random variables in $[0, 1]$, possesses a *completely mixed strategy* (ie, assigning positive probability to all possible types) which is an ESS. Karlin [6] had already reported experimental evidence that the stability condition is far more restrictive than the feasibility condition in this case wrt[1] the existence of ESS (the superiority condition becomes vague in this case since we refer to completely mixed strategies).

Kingman [7] also worked on the severity of the stability condition, in a work on the size of polymorphisms which, interpreted in random evolutionary games, corresponds to random payoff matrices that are *symmetric*. For the case of uniform distribution, he proved that almost all ESS in a random evolutionary game with symmetric payoff matrix, have support size *less than* $2.49\sqrt{n}$. Consequently, Haigh [5] extended this result to the case of asymmetric random payoff matrices. Namely, for a particular probability measure with density $\phi(x) = \exp(-x)/\sqrt{\pi x}$, he proved that almost all ESS have support size *at most* $1.636n^{2/3}$. He also conjectured that similar results should also hold for a wide range of probability measures with continuous density functions.

Another (more recent) line of research concerns the expected number of Nash equilibria in random bimatrix games. Initially McLennan [10] studied this quantity for arbitrary normal form games and provided a formula for this number. Consequently, McLennan and Berg [11] computed asymptotically tight bounds for this formula, for the special case of bimatrix games. They proved that the expected number of NE in normal–random bimatrix games is asymptotically equal to $\exp(0.2816n + \mathcal{O}(\log n))$, while almost all NE have support sizes that concentrate around $0.316n$. Recently Roberts [13] calculated this number in the case zero sum games $\langle A, -A \rangle$ and coordination games $\langle A, A \rangle$, when the Cauchy probability measure is used for the entries of the payoff matrix.

In a previous work of ours [8] we had attempted to study the support sizes of ESS in random games, under the uniform probability measure. In that work we had calculated an exponentially small upper bound on the probability of any given support of size r being the support of an ESS (actually, being the support of a submatrix that satisfies the stability condition). Nevertheless, this bound proved to be insufficient for answering Haigh's conjecture for the uniform case, due to the extremely large number of supports.

In this work we resolve affirmatively the conjecture of Haigh for the cases of both the uniform distribution in $[0, 1]$ and the standard normal distribution (actually, shifted by a positive number). In both cases we prove the crucial probability

---

[1] With respect to.

of stability for a given support to be significantly smaller than exponential. This is enough to prove that almost all ESS have *sublinear* support size. We then proceed to combine our result on satisfaction of the stability condition in random evolutionary games wrt the (standard, shifted) normal distribution, with the result of McLennan and Berg [11] on the expected number of NE in normal–random bimatrix games. Our observation is that ESS in random evolutionary games are significantly less than SNE in the underlying symmetric bimatrix games.

The structure of the rest of the paper is the following: In Section 2 we provide some notation and some elementary background on (symmetric) bimatrix and evolutionary games. In Section 3 we calculate the probability of the stability condition holding (unconditionally) for given support sizes, in the case of the uniform distribution (cf. Subsection 3.2) and in the case of the normal distribution (cf. Subsection 3.3). We then use these bounds to give concentration results on the support sizes of ESS for these two random models of evolutionary games (cf. Subsection 3.4). In Section 4 we prove that the stability condition is more severe than the (symmetric) Nash property in symmetric games, by showing that the expected number of ESS in a evolutionary game with a normal–random payoff matrix A is significantly less than the expected number of Symmetric Nash Equilibria in the underlying symmetric bimatrix game $\langle A, A^T \rangle$.

2   Preliminaries

Notation. $\mathbb{R}$ denotes the set of real numbers, $\mathbb{R}_{\geq 0}$ is the set of nonnegative reals, and $\mathbb{N}$ is the set of nonnegative integer numbers. For any $k \in \mathbb{N} \setminus \{0\}$, we denote the set $\{1, 2, \ldots, k\}$ by $[k]$. $\mathbf{e_i} \in \mathbb{R}^n$ is the vector with all its elements equal to zero, except for its i–th element, which is equal to one. $\mathbf{1} = \sum_{i \in [n]} \mathbf{e_i}$ is the all–one vector, while $\mathbf{0}$ is the all–zero vector in $\mathbb{R}^n$.

We consider any $n \times 1$ matrix as a column vector and any $1 \times n$ matrix as a row vector of $\mathbb{R}^n$. A vector is denoted by small boldface letters (eg, $\mathbf{x}, \mathbf{p}, \ldots$) and is typically considered as a column vector. For any $m \times n$ matrix $A \in \mathbb{R}^{m \times n}$, its i–th row (as a row vector) is denoted by A^i and its j–th column (as a column vector) is denoted by A_j. The (i, j)–th element of A is denoted by $A_{i,j}$ (or, A_{ij}). A^T is the transpose matrix of A. For any positive integer $k \in \mathbb{N}$, $\Delta_k \equiv \{\mathbf{z} \in \mathbb{R}_{\geq 0}^k : \mathbf{1}^T \mathbf{z} = 1\}$ is the $(k-1)$–simplex, ie, the set of probability vectors over k–element sets. For any $\mathbf{z} \in \Delta_k$, its **support** is the subset of $[k]$ of actions that are assigned positive probability mass: $supp(\mathbf{x}) \equiv \{i \in [k] : z_i > 0\}$.

For any probability space $(\Omega, \mathcal{F}, \mathbb{P})$ and any event $\mathcal{E} \in \mathcal{F}$, $\mathbb{P}\{\mathcal{E}\}$ is the probability of this event occurring, while $\mathbb{I}_{\{\mathcal{E}\}}$ is the indicator variable of $\mathcal{E}$ holding. For a random variable X, $\mathbb{E}\{X\}$ is its expectation and $\mathbb{V}ar\{X\}$ its variance. In order to denote that a random variable X gets its value according to a probability distribution F, we use the following notation: $X \in_R F$. For example, for a uniform random variable in $[0, 1]$ we write $X \in_R \mathcal{U}(0, 1)$, while for a random variable drawn from the standard normal distribution we write $X \in_R \mathcal{N}(0, 1)$.

Bimatrix Games. The subclass of symmetric bimatrix games provides the basic setting for much of Evolutionary Game Theory. Indeed, every evolutionary game implies an underlying symmetric bimatrix game, that is repeatedly played between randomly chosen opponents from the population. Therefore we provide the main game theoretic definitions with respect to symmetric bimatrix games.

Definition 1. *For arbitrary $m \times n$ real matrices $A, B \in \mathbb{R}^{m \times n}$, the* **bimatrix game** *$\Gamma = \langle A, B \rangle$ is a game in strategic form between two players, in which the first (row) player has m possible actions and the second (column) player has n possible actions. A* **mixed strategy** *for the row (column) player is a probability distribution $\mathbf{x} \in \Delta_m$ ($\mathbf{y} \in \Delta_n$), according to which she chooses her own action, independently of the other player's choice. A strategy $\mathbf{x} \in \Delta_m$ is* **completely mixed** *if and only if $supp(\mathbf{x}) = [m]$. The* **payoffs** *of the row and the column player, when the row and column players adopt strategies $\mathbf{e_i}$ and $\mathbf{e_j}$, are A_{ij} and B_{ij} respectively. If the two players adopt the strategies $\mathbf{p} \in \Delta_m$ and $\mathbf{q} \in \Delta_n$, then the* **(expected) payoffs** *of the row and column player are $\mathbf{p}^T A \mathbf{q}$ and $\mathbf{p}^T B \mathbf{q}$ respectively. Some special cases of bimatrix games are the* **zero sum** *($B = -A$), the* **coordination** *($B = A$), and the* **symmetric** *($B = A^T$) games.*

Note that in case of a symmetric bimatrix game, the two players have exactly the same set of possible actions (say, $[n]$). The standard notion of equilibrium in strategic games are the Nash Equilibria [12]:

Definition 2. *For any bimatrix game $\langle A, B \rangle$, a strategy profile $(\mathbf{x}, \mathbf{y}) \in \Delta_m \times \Delta_n$ is called a* **Nash Equilibrium** *(NE in short), if and only if $\mathbf{x}^T A \mathbf{y} \geqslant \mathbf{z}^T A \mathbf{y}, \forall \mathbf{z} \in \Delta_m$ and $\mathbf{x}^T B \mathbf{y} \geqslant \mathbf{x}^T B \mathbf{z}, \forall \mathbf{z} \in \Delta_n$. If additionally $supp(\mathbf{x}) = [m]$ and $supp(\mathbf{y}) = [n]$, then $(\mathbf{x}, \mathbf{y})$ is called a* **completely mixed Nash Equilibrium** *(CMNE in short). A profile $(\mathbf{x}, \mathbf{x})$ that is NE for $\langle A, B \rangle$ is called a* **symmetric Nash Equilibrium** *(SNE in short).*

Observe that the payoff matrices in a symmetric bimatrix game need not be symmetric. Note also that not all NE of a symmetric bimatrix game need be symmetric. However it is known that there is at least one such equilibrium:

Theorem 1 ([12]). *Each finite symmetric bimatrix game has at least one SNE.*

When we wish to argue about the vast majority of symmetric bimatrix games, one way is to assume that the real numbers in the set $\{A_{i,j} : (i,j) \in [n]\}$ are independently drawn from a probability distribution F. For example, it can be the uniform distribution in an interval $[a, b] \in \mathbb{R}$, denoted by $\mathcal{U}(a, b)$. Then, a random symmetric bimatrix game Γ is just an instance of the implied random experiment that is described in the following definition:

Definition 3. *A symmetric bimatrix game $\Gamma = \langle A, A^T \rangle$ is an instance of a (symmetric 2-player) random game wrt the probability distribution F, if and only if $\forall i, j \in [n]$, the real number $A_{i,j}$ is an independently and identically distributed random variable drawn from F.*

Evolutionary Stable Strategies. For some $A \in \mathbb{R}^{n \times n}$, fix a symmetric game $\Gamma = \langle A, A^T \rangle$. Suppose that all the individuals of an infinite population are programmed to play the same (either pure or mixed) *incumbent* strategy $\mathbf{x} \in \Delta_n$, whenever they are involved in Γ. Suppose also that at some time a small group of *invaders* appears in the population. Let $\varepsilon \in (0,1)$ be the share of invaders in the post–entry population. Assume that all the invaders are programmed to play the (pure or mixed) strategy $\mathbf{y} \in \Delta_n$ whenever they are involved in Γ.

Pairs of individuals in this *dimorphic* post–entry population are now repeatedly drawn at random to play always the same symmetric game Γ against each other. Recall that, due to symmetry, it is exactly the same for each player to be either the row or the column player. If an individual is chosen to participate, the probability that her (random) opponent will play strategy $\mathbf{x}$ is $1 - \varepsilon$, while that of playing strategy $\mathbf{y}$ is ε. This is equivalent to saying that the opponent is an individual who plays the *mixed* strategy $\mathbf{z} = (1-\varepsilon)\mathbf{x}+\varepsilon\mathbf{y}$. The post–entry payoff to the incumbent strategy $\mathbf{x}$ is then $\mathbf{x}^T A\mathbf{z}$ and that of the invading strategy $\mathbf{y}$ is just $\mathbf{y}^T A\mathbf{z}$. Intuitively, evolutionary forces will select *against* the invader if $\mathbf{x}^T A\mathbf{z} > \mathbf{y}^T A\mathbf{z}$. The most popular notion of stability in evolutionary games is the Evolutionary Stable Strategy (ESS):

Definition 4. *A strategy* $\mathbf{x}$ *is* **evolutionary stable** *(ESS in short) if for any strategy* $\mathbf{y} \neq \mathbf{x}$ *there exists a barrier* $\bar{\varepsilon} = \bar{\varepsilon}(\mathbf{y}) \in (0,1)$ *such that* $\forall 0 < \varepsilon \leqslant \bar{\varepsilon}$, $\mathbf{x}^T A\mathbf{z} > \mathbf{y}^T A\mathbf{z}$ *where* $\mathbf{z} = (1 - \varepsilon)\mathbf{x} + \varepsilon\mathbf{y}$.

The following lemma states that the "hard cases" of evolutionary games are *not* the ones in which there exists a completely mixed ESS:

Lemma 1 (Haigh 1975 [4]). *If a completely mixed strategy* $\mathbf{x} \in \Delta$ *is an ESS, then it is the* unique *ESS of the evolutionary game.*

Indeed, it is true that, if for an evolutionary game with payoff matrix $A \in \mathbb{R}^{n \times n}$ it holds that some strategy $\mathbf{x} \in \Delta_n$ is an ESS, then no strategy $\mathbf{y} \in \Delta_n$ such that $supp(\mathbf{y}) \subseteq supp(\mathbf{x})$ may be an ESS as well.

Haigh [4] also provided an alternative characterization of ESS in evolutionary games, which is the *conjunction* of the following sentences, and will prove to be very useful for our discussion:

Theorem 2 (Haigh [4]). *A strategy* $\mathbf{p} \in \Delta_n$ *in an evolutionary game with payoff matrix* $A \in \mathbb{R}^{n \times n}$ *is an ESS if and only if the following* necessary and sufficient *conditions simultaneously hold:*

[H1]: Nash Property *There is a constant* $c \in \mathbb{R}$ *such that:*

    [H1.1]: Feasibility $\sum_{j \in supp(\mathbf{p})} A_{ij}p_j = A^i\mathbf{p} = c, \ \forall i \in supp(\mathbf{p})$.

    [H1.2]: Superiority $\sum_{j \in supp(\mathbf{p})} A_{ij}p_j = A^i\mathbf{p} \leqslant c, \ \forall i \notin supp(\mathbf{p})$.

[H2]: Stability $\forall \mathbf{x} \in \mathbb{R}^n$:
    IF $(\mathbf{x} \neq \mathbf{0} \wedge supp(\mathbf{x}) \subseteq supp(\mathbf{p}) \wedge \mathbf{1}^T\mathbf{x} = 0)$ **THEN** $\mathbf{x}^T A\mathbf{x} < 0$

Observe that [H1] assures that $(\mathbf{p}, \mathbf{p})$ is a *symmetric Nash Equilibrium* (SNE) of the underlying symmetric bimatrix game $\langle A, A^T \rangle$. This is because $\forall i, j \in [n], i \in supp(\mathbf{p}) \Rightarrow A^i\mathbf{p} \geqslant A^j\mathbf{p}$ and $\forall i, j \in [n], i \in supp(\mathbf{p}) \Rightarrow \mathbf{p}^T(A^T)_i = \mathbf{p}^T(A^i)^T \geqslant \mathbf{p}^T(A^j)^T = \mathbf{p}^T(A^T)_j$. Since in this work we deal with evolutionary games with *random* payoff matrices (in particular, whose entries behave as independent, identically distributed continuous random variables), we can safely assume that almost surely [H1.2] holds with strict inequality. As for [H2], this is the one that guarantees the stability of the strategy against (sufficiently small) invasions.

3   Probability of Stability

In this section we study the probability of a strategy with support size $r \in [n]$ also being an ESS. In the next section we shall use this to calculate an upper bound on the support sizes of almost all ESS in a random game.

Assume a probability distribution F, whose density function $\phi : \mathbb{R} \mapsto [0,1]$ exists, according to which the random variables $\{A_{ij}\}_{(i,j) \in [n] \times [n]}$ determine their values. We focus on the cases of: (i) the uniform distribution $\mathcal{U}(0,1)$, with density function $\phi_u(x) = \mathbb{I}_{\{x \in [0,1]\}}$ and distribution function $\Phi_u(x) = x \cdot \mathbb{I}_{\{x \in [0,1]\}} + \mathbb{I}_{\{x > 1\}}$, and (ii) the standard normal distribution $\mathcal{N}(0,1)$, with density function $\phi_g(x) = \frac{\exp(-x^2/2)}{\sqrt{2\pi}}$ and distribution function $\Phi_g(x) = \int_{-\infty}^{x} n(t)dt$. Our goal is to study the severity of [H2] for a strategy being an ESS. We follow Haigh's generalization of the interesting approach of Kingman (for random *symmetric* payoff matrices) to the case of asymmetric matrices. Our findings are analogous to those of Haigh [5], who gave the general methodology and then focused on a particular distribution. Here we resolve the cases of uniform distribution and standard normal distributions, which were left open in [5].

3.1   Kingman's Approach

Consider an arbitrary strategy $\mathbf{p} \in \Delta_n$, for which we assume (without loss of generality) that its support is $supp(\mathbf{p}) = [r]$. Since condition [H2] has to hold for any non-zero real vector $\mathbf{x} \in \mathbb{R}^n \setminus \{\mathbf{0}\} : \mathbf{1}^T\mathbf{x} = 0 \wedge supp(\mathbf{x}) \subseteq [r]$, we can also apply it for all vectors $\mathbf{x}(\mathbf{i},\mathbf{j}) = \mathbf{e_i} - \mathbf{e_j} : 1 \leqslant i < j \leqslant r$, as was observed in [7]. This immediately implies the following *necessary condition* for $\mathbf{p}$ being an ESS:

$$\forall 1 \leqslant i < j \leqslant r, \ A_{ij} + A_{ji} > A_{ii} + A_{jj} \tag{1}$$

Mimicking Kingman and Haigh's notation [7,5], we denote by D_I the event that our random matrix A has the property described by inequality (1), if $r = |I|$ and we rearrange the rows and columns of A so that $I = [r]$. As was demonstrated in [5], the probability of this event is expressed by the following form:

$$\mathbb{P}\{D_I\} = \int_{-\infty}^{\infty} \cdots \int_{-\infty}^{\infty} \prod_{1 \leqslant i < j \leqslant r} [1 - G(a_{ii} + a_{jj})] \cdot \prod_{i \in [r]} [\phi(a_{ii})] \ da_{11} \cdots da_{rr} \tag{2}$$

where $G(x) = \int_{-\infty}^{x} g(t)dt$ is the distribution function of any random variable $X_{ij} = A_{ij} + A_{ji} : 1 \leqslant i < j \leqslant r$ (the sum of two **iid** random variables with density function ϕ). Note that the density function g is the *convolution* of f with itself. This formula was studied in [5] for the special case $\phi(x) = \exp(-x)/\sqrt{\pi x}$.

In the next two subsections we do the same for the uniform and (shifted) standard normal distribution. Then we bound the support sizes of almost all ESS in uniformly–random and normal–random evolutionary games.

3.2   The Case of $\mathcal{U}(0,1)$

If we adopt $\mathcal{U}(0,1)$ as our basic probability distribution, then of course $f(x) = \phi_u(x) = \mathbb{I}_{\{x \in (0,1)\}}$ and the distribution function G can be easily computed: $\forall 0 \leqslant x \leqslant 1,\ G(x) = \int_0^x f(a_{ii}) \left(\int_0^{x-a_{ii}} f(a_{jj}) da_{jj} \right) da_{ii} = \frac{x^2}{2}$ and $\forall 1 \leqslant x \leqslant 2,\ G(x) = \int_0^1 f(a_{ii}) \left(\int_0^{\min\{1, x-a_{ii}\}} f(a_{jj}) da_{jj} \right) da_{ii} = 2x - 1 - \frac{x^2}{2}$. Therefore we conclude that the following holds (also mentioned in [5]): $\forall x \in \mathbb{R}$,

$$1 - G(x) = \left(1 - \frac{x^2}{2} \right) \cdot \mathbb{I}_{\{0 \leqslant x \leqslant 1\}} + \frac{1}{2}(2-x)^2 \cdot \mathbb{I}_{\{1 < x \leqslant 2\}} \tag{3}$$

Observe now that each $1 - G(a_{ii} + a_{jj})$ factor in equation (2) expresses the probability that the random variable $X_{ij} \equiv A_{ij} + A_{ji}$ is strictly larger than a certain value $a_{ii} + a_{jj}$. On the other hand, all the $f(a_{ii}) = \phi_u(a_{ii})$ factors in equality (2) assure that each of the diagonal elements in A (ie, the random variables A_{ii}) get the assumed values (ie, $A_{ii} = a_{ii}$), which have to be *nonnegative*. We use the following trivial upper bound on each of the $1 - G(a_{ii} + a_{jj})$ factors, which exploits only the fact of *non negative* values of the elementary random variables $A_{ij} \in_R \mathcal{U}(0,1)$ that we consider: $\forall 1 \leqslant i < j \leqslant r,\ 1 - G(a_{ii} + a_{jj}) = \mathbb{P}\{X_{ij} > a_{ii} + a_{jj}\} \leqslant \mathbb{P}\{X_{ij} > a_{ii}\} = 1 - G(a_{ii})$, to get the following from (2):

$$\mathbb{P}\{D_I\} \leqslant \int_0^1 \cdots \int_0^1 \prod_{1 \leqslant i < j \leqslant r} [1 - G(a_{ii})] \ da_{11} \cdots da_{rr}$$

$$= \prod_{i \in [r-1]} \left(\int_0^1 [1 - G(a_{ii})]^{r-i} \ da_{ii} \right) \tag{4}$$

using the facts that $f(x) = \mathbb{I}_{\{x \in (0,1)\}}$ and $\int_0^1 f(a_{rr}) da_{rr} = 1$. Plugging in the form of $1 - G(x)$ in case of the uniform distribution (eq. (3)), we get the following:

$$\mathbb{P}\{D_I\} \leqslant \prod_{i \in [r-1]} \left(\int_0^1 \left[1 - \frac{1}{2} a_{ii}^2 \right]^{r-i} \ da_{ii} \right) \tag{5}$$

Using the trivial bound $(1-x)^a \leqslant \exp(-ax), \forall x > 0, \forall a \geqslant 1$, we have:

$$\mathbb{P}\{D_I\} \leqslant \prod_{i \in [r-1]} \left[\int_0^1 \exp\left(-\frac{r-i}{2} a_{ii}^2 \right) \ da_{ii} \right]$$

$$\leqslant \prod_{i \in [r-1]} \left[\frac{1}{\sqrt{r-i}} \cdot \int_0^1 \exp\left(-\frac{\left(a_{ii}\sqrt{r-i} \right)^2}{2} \right) \ \sqrt{r-i} \ da_{ii} \right]$$

$$= \prod_{i \in [r-1]} \left[\frac{1}{\sqrt{r-i}} \cdot \int_0^{\sqrt{r-i}} \exp\left(-\frac{\beta_i^2}{2}\right) \, d\beta_i \right]$$

$$< \prod_{i \in [r-1]} \left[\frac{1}{2\sqrt{r-i}} \right] = \exp\left(-(r-1)\ln 2 - \frac{1}{2} \sum_{j=1}^{r-1} \ln j \right)$$

$$< \exp\left(-(r-1)\ln 2 - \frac{1}{2} \left[(r-1) + \sum_{j=1}^{r-1} H_j \right] \right)$$

$$= \exp\left(-(r-1)\ln 2 - \frac{1}{2} \left[-2(r-1) + r H_{r-1} \right] \right)$$

$$= \exp\left((r-1)(1 - \ln 2) - \frac{r}{2} - \frac{r}{2}\ln(r-1) \right) = \exp\left(-\frac{r \ln r}{2} + \mathcal{O}(r) \right) \quad (6)$$

since, $\int_0^{\sqrt{r-i}} \exp\left(-\frac{\beta_i^2}{2}\right) d\beta_i < \int_0^\infty \exp\left(-\frac{\beta_i^2}{2}\right) d\beta_i = \frac{1}{2}$. We used the following properties of harmonic numbers: If $H_{r-1} = \sum_{i=1}^{r-1} \frac{1}{i}$ is the $(r-1)$-th harmonic number, then $\sum_{i=1}^{r-1} H_i = r H_{r-1} - (r-1)$ and $\ln(r-1) < H_{r-1} < \ln(r-1) + 1$.

3.3 The Case of $\mathcal{N}(\xi, 1)$

Assume now, for some $\xi > 0$ that will be fixed later, that each element of the payoff matrix behaves as a normally distributed independent random variable with mean ξ and variance 1: $\forall (i,j) \in [n] \times [n], A_{ij} \in_R \mathcal{N}(\xi, 1)$. Then it also holds that all the X_{ij} variables (for $1 \leqslant i < j \leqslant r$) behave also as normally distributed random variables, with mean 2ξ and variance 2. That is: $\forall (i,j) \in [r] \times [r] : i \neq j, X_{ij} \in_R \mathcal{N}(2\xi, 2)$. Then the following hold: $\forall t \in \mathbb{R}, f(t) = \frac{1}{\sqrt{2\pi}} \exp\left(-\frac{(t-\xi)^2}{2}\right)$ and $g(t) = \frac{1}{2\sqrt{\pi}} \exp\left(-\frac{(t-2\xi)^2}{4}\right)$. Moreover, $\forall x \in \mathbb{R}, 1 - F(x) = \frac{1}{\sqrt{2\pi}} \int_x^\infty \exp\left(-\frac{(t-\xi)^2}{2}\right) dt \Rightarrow 1 - F(x) = \frac{1}{\sqrt{2\pi}} \int_{x-\xi}^\infty \exp\left(-\frac{z^2}{2}\right) dz$ (by the change in variable $z = t - \xi$) and $1 - G(x) = \frac{1}{2\sqrt{\pi}} \int_x^\infty \exp\left(-\frac{(t-2\xi)^2}{4}\right) dt \Rightarrow 1 - G(x) = \frac{1}{\sqrt{2\pi}} \int_{\frac{x-2\xi}{\sqrt{2}}}^\infty \exp\left(-\frac{z^2}{2}\right) dz$ (by setting $z = \frac{t-2\xi}{\sqrt{2}}$). The following property is useful for bounding the distribution function of a normal random variable (cf. Theorem 1.4 of [1]): $\forall x > 0, \left(1 - x^{-2}\right) \frac{\exp(x^2/2)}{x} \leqslant \int_x^\infty \exp(-z^2/2) dz \leqslant \frac{\exp(x^2/2)}{x}$. A simple corollary of this property is the following:

Corollary 1. *Assume that $F(x), G(x)$ are the distribution functions of $\mathcal{N}(\xi, 1)$ and $\mathcal{N}(2\xi, 2)$ respectively. Then: $\forall x > \xi, 1 - F(x) \in \left[\left(1 - \frac{1}{(x-\xi)^2}\right), 1 \right] \cdot \frac{1}{\sqrt{2\pi}} \cdot \frac{\exp(-(x-\xi)^2/2)}{x-\xi}$ and $\forall x > 2\xi, 1 - G(x) \in \left[\left(1 - \frac{2}{(x-2\xi)^2}\right), 1 \right] \cdot \frac{1}{\sqrt{\pi}} \cdot \frac{\exp(-(x-2\xi)^2/4)}{x-2\xi}$.*

Recall now that

$$\mathbb{P}\{D_I\} = \underbrace{\int_{-\infty}^\infty \cdots \int_{-\infty}^\infty}_{r \text{ times}} \prod_{1 \leqslant i < j \leqslant r} [1 - G(a_{ii} + a_{jj})] \cdot \prod_{i \in [r]} (f(a_{ii}) da_{ii})$$

$$\leqslant \sum_{k=0}^{r} \binom{r}{k} \prod_{i=1}^{k} \left(\int_{-\infty}^{0} f(a_{ii})da_{ii} \right) \cdot \prod_{i=k+1}^{r} \left(\int_{0}^{\infty} [1 - G(a_{ii})]^{r-i} f(a_{ii})da_{ii} \right)$$

$$= \sum_{k=0}^{r} \binom{r}{k} (F(0))^{k} \cdot \prod_{i=0}^{r-k-1} \mu_i$$

where we have exploited the facts that $\forall x \in \mathbb{R}, 1 - G(x) = \mathbb{P}\{X > x\} \leqslant 1$ and $\forall y, z \geqslant 0, 1 - G(y + z) = \mathbb{P}\{X > y + z\} \leqslant \mathbb{P}\{X > z\} = 1 - G(z)$ and we set $\mu_i \equiv \int_0^\infty [1 - G(x)]^i f(x)dx, \ \forall i \in \mathbb{N}$. Exploiting Corollary 1 and the symmetry of the normal distribution, we have: $F(0)^k = (1 - F(2\xi))^k \leqslant \left(\frac{\exp(-\xi^2/2)}{\xi \cdot \sqrt{2\pi}} \right)^k = \exp\left(-\frac{k \ln(2\pi)}{2} - \frac{k\xi^2}{2} - k \ln \xi \right)$. As for the product of the μ_i's, since $\forall i \geqslant 0, \ \mu_i = \int_0^\infty [1 - G(x)]^i f(x)dx \leqslant [1 - G(0)]^i \cdot (1 - F(0))$, we conclude that: $\prod_{i=0}^{r-k-1} \mu_i \leqslant (1 - F(0))^{r-k}(1 - G(0))^{(r-k)(r-k-1)/2} < \exp\left(-\frac{(r-k)(r-k-1)G(0)}{2} \right)$. Therefore we get the following bound:

$$\mathbb{P}\{D_I\} \leqslant \exp\left(-\frac{r(r-1)}{2} \cdot G(0) \right)$$

$$+ \sum_{k=1}^{r} \exp\left(k \ln\left(\frac{r}{k} \right) - \frac{(r-k)(r-k-1)}{2} \cdot G(0) - \frac{k\xi^2}{2} - \mathcal{O}(k \ln \xi) \right) \quad (7)$$

Assume now that, for some sufficiently small $\delta > 0$, it holds that $\xi = \sqrt{(1-\delta)\ln r}$. Observe that for some constant $\varepsilon > 0$ and all $0 \leqslant k \leqslant \varepsilon r$, $\prod_{i=0}^{r-k-1} \mu_i < \exp\left(-\frac{(1-\varepsilon)^2}{2} r^2 \cdot G(0) \right) = \exp\left(-\frac{(1-\varepsilon)^2}{2} r \ln r \cdot e^{\delta \ln r - \mathcal{O}(\ln \ln r)} \right) < \exp\left(-\frac{(1-\varepsilon)^2}{2} r \ln r \right)$ for $\delta = \Omega\left(\frac{\ln \ln r}{\ln r} \right)$, exploiting the fact that $G(0) = \exp\left(-\xi^2 - \ln \xi - \mathcal{O}(1) \right)$ (cf. Corollary 1). On the other hand, for all $\varepsilon r < k \leqslant r$, observe that $F(0)^k \leqslant \exp\left(-\frac{k\xi^2}{2} - k \ln \xi - \mathcal{O}(k) \right) < \exp\left(-\frac{1-\delta}{2} \varepsilon r \ln r - \mathcal{O}(k \ln \ln r) \right) < \exp\left(-\frac{(1-\delta)\varepsilon}{2} r \ln r \right)$. Since for $\varepsilon = \frac{3 - \sqrt{5}}{2}$ it holds that $\frac{(1-\varepsilon)^2}{2} \geqslant \frac{(1-\delta)\varepsilon}{2}$, we conclude that each term in the right hand side of inequality (7) is upper bounded by $\exp(-\varepsilon(1-\delta)/2 \cdot r \ln r + \mathcal{O}(r))$ and so we get the following: $\mathbb{P}\{D_I\} \leqslant \exp\left(-\frac{(1-\delta)\varepsilon}{2} \cdot r \ln r + \mathcal{O}(r) \right)$.

3.4   Support Sizes of Almost All ESS

In the previous subsections we calculated upper bounds on the probability $\mathbb{P}\{D_I\}$ of a size–r subset $I \subset [n]$ (say, $I = [r]$) satisfying [H2] (and thus being a candidate support for an ESS), for the cases of $\mathcal{U}(0, 1)$ and $\mathcal{N}\left(\sqrt{(1-\delta)\ln r}, 1 \right)$. We now apply the following counting argument introduced by Kingman and used also by Haigh: Let d_r be the event that there exists a submatrix of the random matrix A, of size *at least* $r \times r$, such that D_I is satisfied. Then the probability of this event occurring is upper by the following formula (cf. [5][eq. 10]): $\forall 1 \leqslant s \leqslant r \leqslant n$, $\mathbb{P}\{d_r\} = \mathbb{P}\{\exists \text{ submatrix with } |I| \geqslant r \text{ s.t. } D_I \text{ holds}\} \leqslant \binom{n}{s} \cdot \mathbb{P}\{D_s\} \cdot \binom{r}{s}^{-1}$. Using Stirling's formula, $k! = \sqrt{2\pi k} \cdot (k/e)^k \cdot (1+\Theta(1/k))$,

where $e = \exp(1)$, we write: $\forall 1 \leqslant s \leqslant r \leqslant n$, $\binom{n}{s} \cdot \binom{r}{s}^{-1} = \left(\frac{n}{r}\right)^{s+1/2} \cdot$

$\left(\frac{n}{n-s}\right)^{n-s} \cdot \left(\frac{r-s}{r}\right)^{r-s} \cdot \left(\frac{r-s}{n-s}\right)^{1/2} \cdot (1+o(1))$. Assume now that $r = An^a > s = Bn\beta$, for some $1 > a > \beta > 0$ and $A \geqslant B$. Then: $\binom{n}{s} \cdot \binom{r}{s}^{-1} =$

$(1+o(1)) \cdot \left(\frac{n^{1-a}}{A}\right)^{Bn^\beta+1/2} \cdot \left(1 - \frac{B}{n^{1-\beta}}\right)^{-n^\beta(n^{1-\beta}-B)} \cdot \left(1 - \frac{B}{An^{a-\beta}}\right)^{n^\beta(An^{a-b}-B)} \cdot$

$\left(\frac{A}{n^{1-a}} \cdot \frac{1-B/(An^{a-\beta})}{1-B/n^{1-\beta}}\right)^{1/2} = \exp\left((1-a)Bn^\beta \ln n + \mathcal{O}(n^\beta)\right)$. We proved for the uniform distribution $\mathcal{U}(0,1)$ that for any subset $I \subseteq [n]$ such that $|I| = An^a$, $\mathbb{P}\{D_I\} = \exp\left(-\frac{Aa}{2}n^a \ln n + \mathcal{O}(n^a)\right)$. Therefore, in this case, $\mathbb{P}\{d_{An^a}\} \leqslant \exp\left[-\left(\frac{a}{2}-1+a\right)Bn^\beta \ln n + \mathcal{O}(n^\beta)\right]$, which tends to zero for all $a > 2/3$.

Similarly, we proved for the normal distribution $\mathcal{N}\left(\sqrt{(1-\delta)\ln(Bn^\beta)}, 1\right)$ that for any $I \subseteq [n] : |I| = Bn^\beta$, $\mathbb{P}\{D_I\} = \exp\left(-\frac{\varepsilon(1-\delta)B\beta}{2} \cdot n^\beta \ln n + \mathcal{O}(n^\beta)\right)$, where $\varepsilon = \frac{3-\sqrt{5}}{2}$. Therefore we conclude that: $\mathbb{P}\{d_{An^a}\} \leqslant \exp\left[-\left(\frac{\varepsilon(1-\delta)\beta}{2}-1+a\right)Bn^\beta \ln n + \mathcal{O}(n^\beta)\right]$, which tends to zero for all $a > \frac{4}{7-\sqrt{5}-(3-\sqrt{5})\delta} \cong 0.8396$, since $\delta = \Theta\left(\frac{\ln\ln n}{\ln n}\right) = o(1)$ (for $n \to \infty$). Thus we conclude with the following theorem concerning the support sizes of ESS in a random evolutionary game:

Theorem 3. *Consider an evolutionary game with a random $n \times n$ payoff matrix A. Fix arbitrary positive constant $\zeta > 0$.*

1. *If $A_{ij} \in_R \mathcal{U}(0,1), \forall(i,j) \in [n] \times [n]$, then, as $n \to \infty$, it holds that: $\mathbb{P}\{\exists \text{ ESS with support size at least } n^{(2+2\zeta)/3}\} \leqslant \exp\left(-\frac{5\zeta}{6} \cdot n^{(2+\zeta)/3} \cdot \ln n + \mathcal{O}(n^{(2+\zeta)/3})\right) \to 0$.*

2. *If $A_{ij} \in_R \mathcal{N}(\xi,1), \forall(i,j) \in [n] \times [n]$, where $\xi = \Theta\left(\sqrt{\ln n}\right)$, then, as $n \to \infty$, it holds that: $\mathbb{P}\{\exists \text{ ESS with support size at least } n^{0.8397+\zeta}\} \leqslant \exp\left(-1.19\zeta \cdot n^{0.8397+\zeta/2} \cdot \ln n + \mathcal{O}(n^{0.8397+\zeta/2})\right) \to 0$.*

Remark: Indeed the above theorem upper bounds the *unconditional* probability of [H2] being satisfied by any submatrix of A that is determined by an index set $I \subseteq [n] : |I| > n^{2/3}$ (for the uniform case) or $|I| > n^{0.8397}$ (for the case of the normal distribution). We adopt the particular presentation for purposes of comparison with the corresponding results of Haigh [5] and Kingman [7].

4 An Upper Bound on the Expected Number of ESS

We now combine our result on the probability of [H2] being satisfied in random evolutionary games wrt $\mathcal{N}(\xi, 1)$, with a result of McLennan and Berg [11] on the expected number of NE in random bimatrix games wrt $\mathcal{N}(0, 1)$. The goal is to show that ESS in random evolutionary games are significantly less than SNE in the underlying symmetric bimatrix games.

We start with some additional notation, that will assist the clearer presentation of the argument. Let A, B be normal–random $n \times n$ (payoff) matrices: $\forall(i, j) \in [n] \times [n], A_{ij}, B_{ij} \in_R \mathcal{N}(\xi, 1)$. $E_{n,r}^{nash}$ is the expected number of NE with support sizes equal to r for both strategies, in $\langle A, B \rangle$. $E_{n,r}^{sym}$ is the expected number of SNE with support sizes equal to r for both strategies, in $\langle A, A^T \rangle$. $E_{n,r}^{ess}$ is the expected number of ESS of support size r, in the random evolutionary game, with payoff matrix A. Finally, $E_{n,r}^{stable}$ is the expected number of strategies with support size r that satisfy property [H2], in the random evolutionary game, with payoff matrix A. We shall prove now the following theorem:

Theorem 4. *If the $n \times n$ payoff matrix A of an evolutionary game is randomly chosen so that each of its elements behaves as an independent $\mathcal{N}(\xi, 1)$ random variable, then it holds that $E_n^{ess} = o(E_n^{sym})$, as $n \to \infty$.*

Proof: First of all we should mention that the concept of Nash Equilibrium is invariant under *affine transformations* of the payoff matrices. Therefore, we may safely assume that the results of [11] on the expected number of NE in $n \times n$ bimatrix games, in which the values of both the payoff matrices are treated as standard normal random variables, are also valid if we shift both the payoff matrices by any positive number ξ (or equivalently, if we consider the normal distribution $\mathcal{N}(\xi, 1)$ for the elements of the payoff matrices). The main theorem of the work of McLennan and Berg concerns $E_{n,r}^{nash}$ in $\langle A, B \rangle$ [2].

In our work we are concerned about $E_{n,r}^{ess}$, the expected number of ESS with support size r, in a random evolutionary game with payoff matrix A. Our purpose is to demonstrate the severity of [H2] (compared to the Nash Property [H1] that must also hold for an ESS), therefore we shall compare the expected number of SNE in $\langle A, A^T \rangle$ with the expected number of ESS in the random evolutionary game with payoff matrix A. Although the main result of [11] concerns arbitrary (probably asymmetric) normal–random bimatrix games, if one adapts their calculations for SNE in symmetric bimatrix games, then one can easily observe that similar concentration results hold for this case as well. The key formula of [11] is the following: $\forall 1 \leqslant r \leqslant n, E_{n,r}^{nash} = \binom{n}{r}^2 \cdot 2^{2-2r} \cdot (R(r-1, n-r))^2$,

where, $R(a, b) = \int_{-\infty}^{\infty} \phi_g(x) \cdot \left(\Phi_g \left(\frac{x}{\sqrt{a+1}} \right) \right)^b dx$ is the probability of e_0 getting a value greater than $\sqrt{a+1}$ times the maximum value among $e_1, \ldots, e_b$, where

---

[2] In such a random game, strategy profiles in which the two player don't have the *same* support sizes, are *not* NE with probability asymptotically equal to one. This is why we only focus on profiles in which both players have the same support size r.

$e_0, e_1, \ldots, e_b \in_R \mathcal{N}(0,1)$. For the case of a random symmetric bimatrix game $\langle A, A^T \rangle$, the proper shape of the formula for SNE in $\langle A, A^T \rangle$ is the following:

$$\forall 1 \leqslant r \leqslant n, E_{n,r}^{sym} = \binom{n}{r} \cdot 2^{1-r} \cdot (R(r-1, n-r)).$$

As for the asymptotic result that the support sizes r of NE are sharply concentrated around $0.316n$, this is also valid for SNE in symmetric games. The only difference is that as one increases n by one, the expected number of NE in the symmetric game goes up, *not* by an asymptotic factor of $\exp(0.2816) \approx 1.3252$, but rather by its square root $\exp(0.1408) \approx 1.1512$. So, we can state this extension of the McLennan-Berg result as follows: There exists a constant $\beta \approx 0.316$, such that for any $\varepsilon > 0$, it holds (as $n \rightarrow \infty$) that $\sum_{r=\lfloor (1-\varepsilon)\beta n \rfloor}^{\lceil (1+\varepsilon)\beta n \rceil} E_{n,r}^{sym} \geqslant \varepsilon E_n^{sym}$. From this we can easily deduce that $\sum_{r=1}^{\lfloor (1-\varepsilon)\beta n \rfloor - 1} E_{n,r}^{sym} \leqslant (1-\varepsilon)E_n^{sym}$. It is now rather simple to observe that for any $1 \leqslant Z \leqslant n$, $E_n^{ess} \equiv \sum_{r=1}^{n} E_{n,r}^{ess} = \sum_{r=1}^{Z} E_{n,r}^{ess} + \sum_{r=Z+1}^{n} E_{n,r}^{ess} \leqslant \sum_{r=1}^{Z} E_{n,r}^{sym} + \sum_{r=Z+1}^{n} E_{n,r}^{stable}$, since $ess = stable \wedge sym$. If we set $Z = n^{0.8397+\zeta}$ for some $\zeta > 0$, then: $\sum_{r=1}^{Z} E_{n,r}^{sym} \leqslant \frac{n^{-0.1603+\zeta}}{\beta} E_n^{sym}$ and $\sum_{r=Z+1}^{n} E_{n,r}^{stable} \leqslant \sum_{r=Z+1}^{n} E_n^{ess}$. $\mathbb{P}\{\exists \text{ ESS with support } \geqslant r\} < \sum_{Z=r+1}^{n} E_n^{sym} \cdot \mathbb{P}\{\exists \text{ ESS with support } \geqslant r\} < E_n^{sym} \cdot \exp\left(\log n - 1.19\zeta \cdot n^{0.8397+\zeta/2} \cdot \ln n + \mathcal{O}\left(n^{0.8397+\zeta/2}\right)\right)$. Therefore, we conclude that $E_n^{ess} = \mathcal{O}\left(n^{-0.16} \cdot E_n^{sym}\right) = o(E_n^{sym})$. ∎

References

1. Durrett, R.: Probability: Theory and Examples, 2nd edn. Duxbury Press (1996)
2. Etessami, K., Lochbihler, A.: The computational complexity of evolutionary stable strategies. Int. J. of Game Theory (to appear, 2007)
3. Gillespie, J.H.: A general model to account for enzyme variation in natural populations. III: Multiple alleles. Evolution 31, 85–90 (1977)
4. Haigh, J.: Game theory and evolution. Adv. in Appl. Prob. 7, 8–11 (1975)
5. Haigh, J.: How large is the support of an ESS? J. of Appl. Prob. 26, 164–170 (1989)
6. Karlin, S.: Some natural viability systems for a multiallelic locus: A theoretical study. Genetics 97, 457–473 (1981)
7. Kingman, J.F.C.: Typical polymorphisms maintained by selection at a single locus (special issue). J. of Appl. Prob. 25, 113–125 (1988)
8. Kontogiannis, S., Spirakis, P.: Counting stable strategies in random evolutionary games. In: Deng, X., Du, D.-Z. (eds.) ISAAC 2005. LNCS, vol. 3827, pp. 839–848. Springer, Heidelberg (2005)
9. Lewontin, R.C., Ginzburg, L.R., Tuljapurkar, S.D.: Heterosis as an explanation of large amounts of genic polymorphism. Genetics 88, 149–169 (1978)
10. McLennan, A.: The expected numer of nash equilibria of a normal form game. Econometrica 73(1), 141–174 (2005)
11. McLennan, A., Berg, J.: The asymptotic expected number of nash equilibria of two player normal form games. Games & Econ. Behavior 51, 264–295 (2005)
12. Nash, J.: Noncooperative games. Annals of Mathematics 54, 289–295 (1951)
13. Roberts, D.P.: Nash equilibria of cauchy-random zero-sum and coordination matrix games. Int. J. of Game Theory 34, 167–184 (2006)
14. Smith, J.M., Price, G.: The logic of animal conflict. Nature 246, 15–18 (1973)

Author Index

Ackermann, Heiner   58

Buhrman, Harry   1

Czumaj, Artur   70

Dietzfelbinger, Martin   2

Erlebach, Thomas   82

Freivalds, Rūsiņš   18

Gutjahr, Walter J.   93

Hall, Alexander   82

Jägersküpper, Jens   118
Jao, David   105

Kapoutsis, Christos   130
Karakoc, Mustafa   142

Katzensteiner, Stefan   93
Kavak, Adnan   142
Kontogiannis, Spyros C.   30, 154
Královič, Richard   130

Mihal'ák, Matúš   82
Mömke, Tobias   130

Raju, S. Ramesh   105
Reiter, Peter   93

Spirakis, Paul G.   30, 154
Srinivasan, Aravind   54

Venkatesan, Ramarathnam   105

Wang, Xin   70

Lecture Notes in Computer Science

Sublibrary 1: Theoretical Computer Science and General Issues

For information about Vols. 1– 4422
please contact your bookseller or Springer

Vol. 4743: P. Thulasiraman, X. He, T.L. Xu, M.K. Denko, R.K. Thulasiram, L.T. Yang (Eds.), Frontiers of High Performance Computing and Networking ISPA 2007 Workshops. XXIX, 536 pages. 2007.

Vol. 4742: I. Stojmenovic, R.K. Thulasiram, L.T. Yang, W. Jia, M. Guo, R.F. de Mello (Eds.), Parallel and Distributed Processing and Applications. XX, 995 pages. 2007.

Vol. 4736: S. Winter, M. Duckham, L. Kulik, B. Kuipers (Eds.), Spatial Information Theory. XV, 455 pages. 2007.

Vol. 4732: K. Schneider, J. Brandt (Eds.), Theorem Proving in Higher Order Logics. IX, 401 pages. 2007.

Vol. 4708: L. Kučera, A. Kučera (Eds.), Mathematical Foundations of Computer Science 2007. XVIII, 764 pages. 2007.

Vol. 4707: O. Gervasi, M.L. Gavrilova (Eds.), Computational Science and Its Applications – ICCSA 2007, Part III. XXIV, 1205 pages. 2007.

Vol. 4706: O. Gervasi, M.L. Gavrilova (Eds.), Computational Science and Its Applications – ICCSA 2007, Part II. XXIII, 1129 pages. 2007.

Vol. 4705: O. Gervasi, M.L. Gavrilova (Eds.), Computational Science and Its Applications – ICCSA 2007, Part I. XLIV, 1169 pages. 2007.

Vol. 4703: L. Caires, V.T. Vasconcelos (Eds.), CONCUR 2007 – Concurrency Theory. XIII, 507 pages. 2007.

Vol. 4697: L. Choi, Y. Paek, S. Cho (Eds.), Advances in Computer Systems Architecture. XIII, 400 pages. 2007.

Vol. 4688: K. Li, M. Fei, G.W. Irwin, S. Ma (Eds.), Bio-Inspired Computational Intelligence and Applications. XIX, 805 pages. 2007.

Vol. 4684: L. Kang, Y. Liu, S. Zeng (Eds.), Evolvable Systems: From Biology to Hardware. XIV, 446 pages. 2007.

Vol. 4683: L. Kang, Y. Liu, S. Zeng (Eds.), Intelligence Computation and Applications. XVII, 663 pages. 2007.

Vol. 4681: D.-S. Huang, L. Heutte, M. Loog (Eds.), Advanced Intelligent Computing Theories and Applications. XXVI, 1379 pages. 2007.

Vol. 4671: V. Malyshkin (Ed.), Parallel Computing Technologies. XIV, 635 pages. 2007.

Vol. 4669: J.M. de Sá, L.A. Alexandre, W. Duch, D. Mandic (Eds.), Artificial Neural Networks – ICANN 2007, Part II. XXXI, 990 pages. 2007.

Vol. 4668: J.M. de Sá, L.A. Alexandre, W. Duch, D. Mandic (Eds.), Artificial Neural Networks – ICANN 2007, Part I. XXXI, 978 pages. 2007.

Vol. 4666: M.E. Davies, C.J. James, S.A. Abdallah, M.D Plumbley (Eds.), Independent Component Analysis and Blind Signal Separation. XIX, 847 pages. 2007.

Vol. 4665: J. Hromkovič, R. Královič, M. Nunkesser, P. Widmayer (Eds.), Stochastic Algorithms: Foundations and Applications. X, 167 pages. 2007.

Vol. 4664: J. Durand-Lose, M. Margenstern (Eds.), Machines, Computations, and Universality. X, 325 pages. 2007.

Vol. 4649: V. Diekert, M.V. Volkov, A. Voronkov (Eds.), Computer Science – Theory and Applications. XIII, 420 pages. 2007.

Vol. 4647: R. Martin, M. Sabin, J. Winkler (Eds.), Mathematics of Surfaces XII. IX, 509 pages. 2007.

Vol. 4644: N. Azemard, L. Svensson (Eds.), Integrated Circuit and System Design. XIV, 583 pages. 2007.

Vol. 4641: A.-M. Kermarrec, L. Bougé, T. Priol (Eds.), Euro-Par 2007 Parallel Processing. XXVII, 974 pages. 2007.

Vol. 4639: E. Csuhaj-Varjú, Z. Ésik (Eds.), Fundamentals of Computation Theory. XIV, 508 pages. 2007.

Vol. 4638: T. Stützle, M. Birattari, H.H. Hoos (Eds.), Engineering Stochastic Local Search Algorithms. X, 223 pages. 2007.

Vol. 4628: L.N. de Castro, F.J. Von Zuben, H. Knidel (Eds.), Artificial Immune Systems. XII, 438 pages. 2007.

Vol. 4627: M. Charikar, K. Jansen, O. Reingold, J.D.P. Rolim (Eds.), Approximation, Randomization, and Combinatorial Optimization. XII, 626 pages. 2007.

Vol. 4624: T. Mossakowski, U. Montanari, M. Haveraaen (Eds.), Algebra and Coalgebra in Computer Science. XI, 463 pages. 2007.

Vol. 4619: F. Dehne, J.-R. Sack, N. Zeh (Eds.), Algorithms and Data Structures. XVI, 662 pages. 2007.

Vol. 4618: S.G. Akl, C.S. Calude, M.J. Dinneen, G. Rozenberg, H.T. Wareham (Eds.), Unconventional Computation. X, 243 pages. 2007.

Vol. 4616: A. Dress, Y. Xu, B. Zhu (Eds.), Combinatorial Optimization and Applications. XI, 390 pages. 2007.

Vol. 4613: F.P. Preparata, Q. Fang (Eds.), Frontiers in Algorithmics. XI, 348 pages. 2007.

Vol. 4600: H. Comon-Lundh, C. Kirchner, H. Kirchner (Eds.), Rewriting, Computation and Proof. XVI, 273 pages. 2007.

Vol. 4599: S. Vassiliadis, M. Berekovic, T.D. Hämäläinen (Eds.), Embedded Computer Systems: Architectures, Modeling, and Simulation. XVIII, 466 pages. 2007.

Vol. 4598: G. Lin (Ed.), Computing and Combinatorics. XII, 570 pages. 2007.

Vol. 4596: L. Arge, C. Cachin, T. Jurdziński, A. Tarlecki (Eds.), Automata, Languages and Programming. XVII, 953 pages. 2007.

Vol. 4595: D. Bošnački, S. Edelkamp (Eds.), Model Checking Software. X, 285 pages. 2007.

Vol. 4590: W. Damm, H. Hermanns (Eds.), Computer Aided Verification. XV, 562 pages. 2007.

Vol. 4588: T. Harju, J. Karhumäki, A. Lepistö (Eds.), Developments in Language Theory. XI, 423 pages. 2007.

Vol. 4583: S.R. Della Rocca (Ed.), Typed Lambda Calculi and Applications. X, 397 pages. 2007.

Vol. 4580: B. Ma, K. Zhang (Eds.), Combinatorial Pattern Matching. XII, 366 pages. 2007.

Vol. 4576: D. Leivant, R. de Queiroz (Eds.), Logic, Language, Information and Computation. X, 363 pages. 2007.

Vol. 4547: C. Carlet, B. Sunar (Eds.), Arithmetic of Finite Fields. XI, 355 pages. 2007.

Vol. 4546: J. Kleijn, A. Yakovlev (Eds.), Petri Nets and Other Models of Concurrency – ICATPN 2007. XI, 515 pages. 2007.

Vol. 4545: H. Anai, K. Horimoto, T. Kutsia (Eds.), Algebraic Biology. XIII, 379 pages. 2007.

Vol. 4533: F. Baader (Ed.), Term Rewriting and Applications. XII, 419 pages. 2007.

Vol. 4528: J. Mira, J.R. Álvarez (Eds.), Nature Inspired Problem-Solving Methods in Knowledge Engineering, Part II. XXII, 650 pages. 2007.

Vol. 4527: J. Mira, J.R. Álvarez (Eds.), Bio-inspired Modeling of Cognitive Tasks, Part I. XXII, 630 pages. 2007.

Vol. 4525: C. Demetrescu (Ed.), Experimental Algorithms. XIII, 448 pages. 2007.

Vol. 4514: S.N. Artemov, A. Nerode (Eds.), Logical Foundations of Computer Science. XI, 513 pages. 2007.

Vol. 4513: M. Fischetti, D.P. Williamson (Eds.), Integer Programming and Combinatorial Optimization. IX, 500 pages. 2007.

Vol. 4510: P. Van Hentenryck, L.A. Wolsey (Eds.), Integration of AI and OR Techniques in Constraint Programming for Combinatorial Optimization Problems. X, 391 pages. 2007.

Vol. 4507: F. Sandoval, A. Prieto, J. Cabestany, M. Graña (Eds.), Computational and Ambient Intelligence. XXVI, 1167 pages. 2007.

Vol. 4501: J. Marques-Silva, K.A. Sakallah (Eds.), Theory and Applications of Satisfiability Testing – SAT 2007. XI, 384 pages. 2007.

Vol. 4497: S.B. Cooper, B. Löwe, A. Sorbi (Eds.), Computation and Logic in the Real World. XVIII, 826 pages. 2007.

Vol. 4494: H. Jin, O.F. Rana, Y. Pan, V.K. Prasanna (Eds.), Algorithms and Architectures for Parallel Processing. XIV, 508 pages. 2007.

Vol. 4493: D. Liu, S. Fei, Z. Hou, H. Zhang, C. Sun (Eds.), Advances in Neural Networks – ISNN 2007, Part III. XXVI, 1215 pages. 2007.

Vol. 4492: D. Liu, S. Fei, Z. Hou, H. Zhang, C. Sun (Eds.), Advances in Neural Networks – ISNN 2007, Part II. XXVII, 1321 pages. 2007.

Vol. 4491: D. Liu, S. Fei, Z.-G. Hou, H. Zhang, C. Sun (Eds.), Advances in Neural Networks – ISNN 2007, Part I. LIV, 1365 pages. 2007.

Vol. 4490: Y. Shi, G.D. van Albada, J. Dongarra, P.M.A. Sloot (Eds.), Computational Science – ICCS 2007, Part IV. XXXVII, 1211 pages. 2007.

Vol. 4489: Y. Shi, G.D. van Albada, J. Dongarra, P.M.A. Sloot (Eds.), Computational Science – ICCS 2007, Part III. XXXVII, 1257 pages. 2007.

Vol. 4488: Y. Shi, G.D. van Albada, J. Dongarra, P.M.A. Sloot (Eds.), Computational Science – ICCS 2007, Part II. XXXV, 1251 pages. 2007.

Vol. 4487: Y. Shi, G.D. van Albada, J. Dongarra, P.M.A. Sloot (Eds.), Computational Science – ICCS 2007, Part I. LXXXI, 1275 pages. 2007.

Vol. 4484: J.-Y. Cai, S.B. Cooper, H. Zhu (Eds.), Theory and Applications of Models of Computation. XIII, 772 pages. 2007.

Vol. 4475: P. Crescenzi, G. Prencipe, G. Pucci (Eds.), Fun with Algorithms. X, 273 pages. 2007.

Vol. 4474: G. Prencipe, S. Zaks (Eds.), Structural Information and Communication Complexity. XI, 342 pages. 2007.

Vol. 4459: C. Cérin, K.-C. Li (Eds.), Advances in Grid and Pervasive Computing. XVI, 759 pages. 2007.

Vol. 4449: Z. Horváth, V. Zsók, A. Butterfield (Eds.), Implementation and Application of Functional Languages. X, 271 pages. 2007.

Vol. 4448: M. Giacobini (Ed.), Applications of Evolutionary Computing. XXIII, 755 pages. 2007.

Vol. 4447: E. Marchiori, J.H. Moore, J.C. Rajapakse (Eds.), Evolutionary Computation, Machine Learning and Data Mining in Bioinformatics. XI, 302 pages. 2007.

Vol. 4446: C. Cotta, J. van Hemert (Eds.), Evolutionary Computation in Combinatorial Optimization. XII, 241 pages. 2007.

Vol. 4445: M. Ebner, M. O'Neill, A. Ekárt, L. Vanneschi, A.I. Esparcia-Alcázar (Eds.), Genetic Programming. XI, 382 pages. 2007.

Vol. 4436: C.R. Stephens, M. Toussaint, D. Whitley, P.F. Stadler (Eds.), Foundations of Genetic Algorithms. IX, 213 pages. 2007.

Vol. 4433: E. Şahin, W.M. Spears, A.F.T. Winfield (Eds.), Swarm Robotics. XII, 221 pages. 2007.

Vol. 4432: B. Beliczynski, A. Dzielinski, M. Iwanowski, B. Ribeiro (Eds.), Adaptive and Natural Computing Algorithms, Part II. XXVI, 761 pages. 2007.

Vol. 4431: B. Beliczynski, A. Dzielinski, M. Iwanowski, B. Ribeiro (Eds.), Adaptive and Natural Computing Algorithms, Part I. XXV, 851 pages. 2007.

Vol. 4424: O. Grumberg, M. Huth (Eds.), Tools and Algorithms for the Construction and Analysis of Systems. XX, 738 pages. 2007.

Vol. 4423: H. Seidl (Ed.), Foundations of Software Science and Computational Structures. XVI, 379 pages. 2007.